Apache Kafka Quick Start Guide

I0004053

Leverage Apache Kafka 2.0 to simplify real-time data processing for distributed applications

Raúl Estrada

BIRMINGHAM - MUMBAI

Apache Kafka Quick Start Guide

Copyright © 2018 Packt Publishing

All rights reserved. No part of this book may be reproduced, stored in a retrieval system, or transmitted in any form or by any means, without the prior written permission of the publisher, except in the case of brief quotations embedded in critical articles or reviews.

Every effort has been made in the preparation of this book to ensure the accuracy of the information presented. However, the information contained in this book is sold without warranty, either express or implied. Neither the author, nor Packt Publishing or its dealers and distributors, will be held liable for any damages caused or alleged to have been caused directly or indirectly by this book.

Packt Publishing has endeavored to provide trademark information about all of the companies and products mentioned in this book by the appropriate use of capitals. However, Packt Publishing cannot guarantee the accuracy of this information.

Commissioning Editor: Amey Varangaonkar
Acquisition Editor: Siddharth Mandal
Content Development Editor: Smit Carvalho
Technical Editor: Niral Almeida
Copy Editor: Safis Editing
Project Coordinator: Pragati Shukla
Proofreader: Safis Editing
Indexer: Mariammal Chettiyar
Graphics: Jason Monteiro
Production Coordinator: Deepika Naik

First published: December 2018

Production reference: 1261218

Published by Packt Publishing Ltd.
Livery Place
35 Livery Street
Birmingham
B3 2PB, UK.

ISBN 978-1-78899-782-9

www.packtpub.com

This book is dedicated to my mom, who convinced me to write it, for her sacrifices and for exemplifying the power of self confidence.

– Raúl Estrada

`mapt.io`

Mapt is an online digital library that gives you full access to over 5,000 books and videos, as well as industry leading tools to help you plan your personal development and advance your career. For more information, please visit our website.

Why subscribe?

- Spend less time learning and more time coding with practical eBooks and Videos from over 4,000 industry professionals

- Improve your learning with Skill Plans built especially for you

- Get a free eBook or video every month

- Mapt is fully searchable

- Copy and paste, print, and bookmark content

Packt.com

Did you know that Packt offers eBook versions of every book published, with PDF and ePub files available? You can upgrade to the eBook version at `www.packt.com` and as a print book customer, you are entitled to a discount on the eBook copy. Get in touch with us at `customercare@packtpub.com` for more details.

At `www.packt.com`, you can also read a collection of free technical articles, sign up for a range of free newsletters, and receive exclusive discounts and offers on Packt books and eBooks.

Contributors

About the author

Raúl Estrada has been a programmer since 1996 and a Java developer since 2001. He loves all topics related to computer science. With more than 15 years of experience in high-availability and enterprise software, he has been designing and implementing architectures since 2003. His specialization is in systems integration, and he mainly participates in projects related to the financial sector. He has been an enterprise architect for BEA Systems and Oracle Inc., but he also enjoys web, mobile, and game programming. Raúl is a supporter of free software and enjoys experimenting with new technologies, frameworks, languages, and methods.

Raúl is the author of other Packt Publishing titles, such as *Fast Data Processing Systems with SMACK* and *Apache Kafka Cookbook*.

> *I want to say thanks to the technical reviewer, Isaac Ruiz; without his effort and patience, it would not have been possible to write this book.*

> *I also thank Siddharth Mandal, the acquisition editor, who believed in this project from the beginning.*

> *And finally, I want to thank all the heroes who contribute to open source projects, specifically with Apache Kafka.*

About the reviewer

Isaac Ruiz Guerra has been a Java programmer since 2001 and an IT consultant since 2003. Isaac is specialized in systems integration, and has participated in projects to do with the financial sector. Isaac has worked mainly on the backend side, using languages such as Java, Python, and Elixir. For more than 10 years, he has worked with different application servers for the Java world, including JBoss, Glassfish, and WLS. Isaac is currently interested in topics such as microservices, cloud native, and serverless. He is a regular lecturer, mainly at conferences related to the JVM. Isaac is interested in the formation of interdisciplinary and high-performance teams.

Packt is searching for authors like you

If you're interested in becoming an author for Packt, please visit `authors.packtpub.com` and apply today. We have worked with thousands of developers and tech professionals, just like you, to help them share their insight with the global tech community. You can make a general application, apply for a specific hot topic that we are recruiting an author for, or submit your own idea.

Table of Contents

Preface

Since 2011, Kafka's been exploding in terms of growth. More than one third of Fortune 500 companies use Apache Kafka. These companies include travel companies, banks, insurance companies, and telecom companies.

Uber, Twitter, Netflix, Spotify, Blizzard, LinkedIn, Spotify, and PayPal process their messages with Apache Kafka every day.

Today, Apache Kafka is used to collect data, do real-time data analysis, and perform real-time data streaming. Kafka is also used to feed events to **Complex Event Processing** (**CEP**) architectures, is deployed in microservice architectures, and is implemented in **Internet of Things** (**IoT**) systems.

In the realm of streaming, there are several competitors to Kafka Streams, including Apache Spark, Apache Flink, Akka Streams, Apache Pulsar, and Apache Beam. They are all in competition to perform better than Kafka. However, Apache Kafka has one key advantage over them all: its ease of use. Kafka is easy to implement and maintain, and its learning curve is not very steep.

This book is a practical quick start guide. It is focused on showing practical examples and does not get involved in theoretical explanations or discussions of Kafka's architecture. This book is a compendium of hands-on recipes, solutions to everyday problems faced by those implementing Apache Kafka.

Who this book is for

This book is for data engineers, software developers, and data architects looking for a quick hands-on Kafka guide.

This guide is about programming; it is an introduction for those with no previous knowledge about Apache Kafka.

All the examples are written in Java 8; experience with Java 8 is the only requirement for following this guide.

What this book covers

Chapter 1, *Configuring Kafka*, explains the basics for getting started with Apache Kafka. It discusses how to install, configure, and run Kafka. It also discusses how to make basic operations with Kafka brokers and topics.

Chapter 2, *Message Validation*, explores how to program data validation for your enterprise service bus, covering how to filter messages from an input stream.

Chapter 3, *Message Enrichment*, looks at message enrichment, another important task for an enterprise service bus. Message enrichment is the process of incorporating additional information into the messages of your stream.

Chapter 4, *Serialization*, talks about how to build serializers and deserializers for writing, reading, or converting messages in binary, raw string, JSON, or AVRO formats.

Chapter 5, *Schema Registry*, covers how to validate, serialize, deserialize, and keep a history of versions of messages using the Kafka Schema Registry.

Chapter 6, *Kafka Streams*, explains how to obtain information about a group of messages – in other words, a message stream – and how to obtain additional information, such as that to do with the aggregation and composition of messages, using Kafka Streams.

Chapter 7, *KSQL*, talks about how to manipulate event streams without a single line of code using SQL over Kafka Streams.

Chapter 8, *Kafka Connect*, talks about other fast data processing tools and how to make a data processing pipeline with them in conjunction with Apache Kafka. Tools such as Apache Spark and Apache Beam are covered in this chapter.

To get the most out of this book

The reader should have some experience of programming with Java 8.

The minimum configuration required for executing the recipes in this book is an Intel ® Core i3 Processor, 4 GB of RAM, and 128 GB of disk space. Linux or macOS is recommended, as Windows is not fully supported.

Download the example code files

You can download the example code files for this book from your account at www.packt.com. If you purchased this book elsewhere, you can visit www.packt.com/support and register to have the files emailed directly to you.

You can download the code files by following these steps:

1. Log in or register at www.packt.com.
2. Select the **SUPPORT** tab.
3. Click on **Code Downloads & Errata**.
4. Enter the name of the book in the **Search** box and follow the onscreen instructions.

Once the file is downloaded, please make sure that you unzip or extract the folder using the latest version of:

- WinRAR/7-Zip for Windows
- Zipeg/iZip/UnRarX for Mac
- 7-Zip/PeaZip for Linux

The code bundle for the book is also hosted on GitHub at https://github.com/PacktPublishing/Apache-Kafka-Quick-Start-Guide. In case there's an update to the code, it will be updated on the existing GitHub repository.

We also have other code bundles from our rich catalog of books and videos available at https://github.com/PacktPublishing/. Check them out!

Conventions used

There are a number of text conventions used throughout this book.

CodeInText: Indicates code words in text, database table names, folder names, filenames, file extensions, pathnames, dummy URLs, user input, and Twitter handles. Here is an example: "The --topic parameter sets the name of the topic; in this case, amazingTopic."

A block of code is set as follows:

```
{
    "event": "CUSTOMER_CONSULTS_ETHPRICE",
    "customer": {
        "id": "14862768",
        "name": "Snowden, Edward",
        "ipAddress": "95.31.18.111"
    },
    "currency": {
        "name": "ethereum",
        "price": "RUB"
    },
    "timestamp": "2018-09-28T09:09:09Z"
}
```

When we wish to draw your attention to a particular part of a code block, the relevant lines or items are set in bold:

```
dependencies {
    compile group: 'org.apache.kafka', name: 'kafka_2.12', version:
                                                        '2.0.0'
    compile group: 'com.maxmind.geoip', name: 'geoip-api', version:
                                                        '1.3.1'
    compile group: 'com.fasterxml.jackson.core', name: 'jackson-core',
version: '2.9.7'
}
```

Any command-line input or output is written as follows:

```
> <confluent-path>/bin/kafka-topics.sh --list --ZooKeeper localhost:2181
```

Bold: Indicates a new term, an important word, or words that you see onscreen. For example, words in menus or dialog boxes appear in the text like this. Here is an example: "To differentiate among them, the events on **t1** have one stripe, the events on **t2** have two stripes, and the events on **t3** have three stripes."

Warnings or important notes appear like this.

Tips and tricks appear like this.

Get in touch

Feedback from our readers is always welcome.

General feedback: If you have questions about any aspect of this book, mention the book title in the subject of your message and email us at customercare@packtpub.com.

Errata: Although we have taken every care to ensure the accuracy of our content, mistakes do happen. If you have found a mistake in this book, we would be grateful if you would report this to us. Please visit www.packt.com/submit-errata, selecting your book, clicking on the Errata Submission Form link, and entering the details.

Piracy: If you come across any illegal copies of our works in any form on the Internet, we would be grateful if you would provide us with the location address or website name. Please contact us at copyright@packt.com with a link to the material.

If you are interested in becoming an author: If there is a topic that you have expertise in and you are interested in either writing or contributing to a book, please visit authors.packtpub.com.

Reviews

Please leave a review. Once you have read and used this book, why not leave a review on the site that you purchased it from? Potential readers can then see and use your unbiased opinion to make purchase decisions, we at Packt can understand what you think about our products, and our authors can see your feedback on their book. Thank you!

For more information about Packt, please visit packt.com.

1
Configuring Kafka

This chapter describes what Kafka is and the concepts related to this technology: brokers, topics, producers, and consumers. It also talks about how to build a simple producer and consumer from the command line, as well as how to install Confluent Platform. The information in this chapter is fundamental to the following chapters.

In this chapter, we will cover the following topics:

- Kafka in a nutshell
- Installing Kafka (Linux and macOS)
- Installing the Confluent Platform
- Running Kafka
- Running Confluent Platform
- Running Kafka brokers
- Running Kafka topics
- A command–line message producer
- A command–line message consumer
- Using kafkacat

Kafka in a nutshell

Apache Kafka is an open source streaming platform. If you are reading this book, maybe you already know that Kafka scales very well in a horizontal way without compromising speed and efficiency.

The Kafka core is written in Scala, and Kafka Streams and KSQL are written in Java. A Kafka server can run in several operating systems: Unix, Linux, macOS, and even Windows. As it usually runs in production on Linux servers, the examples in this book are designed to run on Linux environments. The examples in this book also consider bash environment usage.

This chapter explains how to install, configure, and run Kafka. As this is a Quick Start Guide, it does not cover Kafka's theoretical details. At the moment, it is appropriate to mention these three points:

- **Kafka is a service bus**: To connect heterogeneous applications, we need to implement a message publication mechanism to send and receive messages among them. A message router is known as message broker. Kafka is a message broker, a solution to deal with routing messages among clients in a quick way.
- **Kafka architecture has two directives**: The first is to not block the producers (in order to deal with the back pressure). The second is to isolate producers and consumers. The producers should not know who their consumers are, hence Kafka follows the dumb broker and smart clients model.
- **Kafka is a real-time messaging system**: Moreover, Kafka is a software solution with a publish-subscribe model: open source, distributed, partitioned, replicated, and commit-log-based.

There are some concepts and nomenclature in Apache Kafka:

- **Cluster**: This is a set of Kafka brokers.
- **Zookeeper**: This is a cluster coordinator—a tool with different services that are part of the Apache ecosystem.
- **Broker**: This is a Kafka server, also the Kafka server process itself.
- **Topic**: This is a queue (that has log partitions); a broker can run several topics.
- **Offset**: This is an identifier for each message.
- **Partition**: This is an immutable and ordered sequence of records continually appended to a structured commit log.
- **Producer**: This is the program that publishes data to topics.
- **Consumer**: This is the program that processes data from the topics.
- **Retention period**: This is the time to keep messages available for consumption.

In Kafka, there are three types of clusters:

- Single node–single broker
- Single node–multiple broker
- Multiple node–multiple broker

In Kafka, there are three (and just three) ways to deliver messages:

- **Never redelivered**: The messages may be lost because, once delivered, they are not sent again.
- **May be redelivered**: The messages are never lost because, if it is not received, the message can be sent again.
- **Delivered once**: The message is delivered exactly once. This is the most difficult form of delivery; since the message is only sent once and never redelivered, it implies that there is zero loss of any message.

The message log can be compacted in two ways:

- **Coarse-grained**: Log compacted by time
- **Fine-grained**: Log compacted by message

Kafka installation

There are three ways to install a Kafka environment:

- Downloading the executable files
- Using `brew` (in macOS) or `yum` (in Linux)
- Installing Confluent Platform

For all three ways, the first step is to install Java; we need Java 8. Download and install the latest JDK 8 from the Oracle's website:

```
http://www.oracle.com/technetwork/java/javase/downloads/index.html
```

At the time of writing, the latest Java 8 JDK version is 8u191.

For Linux users :

1. Change the file mode to executable as follows, follows these steps:

```
> chmod +x jdk-8u191-linux-x64.rpm
```

2. Go to the directory in which you want to install Java:

```
> cd <directory path>
```

3. Run the `rpm` installer with the following command:

```
> rpm -ivh jdk-8u191-linux-x64.rpm
```

4. Add to your environment the JAVA_HOME variable. The following command writes the JAVA_HOME environment variable to the `/etc/profile` file:

```
> echo "export JAVA_HOME=/usr/java/jdk1.8.0_191" >> /etc/profile
```

5. Validate the Java installation as follows:

```
> java -version
java version "1.8.0_191"
Java(TM) SE Runtime Environment (build 1.8.0_191-b12)
Java HotSpot(TM) 64-Bit Server VM (build 25.191-b12, mixed mode)
```

At the time of writing, the latest Scala version is 2.12.6. To install Scala in Linux, perform the following steps:

1. Download the latest Scala binary from http://www.scala-lang.org/download
2. Extract the downloaded file, `scala-2.12.6.tgz`, as follows:

```
> tar xzf scala-2.12.6.tgz
```

3. Add the SCALA_HOME variable to your environment as follows:

```
> export SCALA_HOME=/opt/scala
```

4. Add the Scala bin directory to your PATH environment variable as follows:

```
> export PATH=$PATH:$SCALA_HOME/bin
```

5. To validate the Scala installation, do the following:

```
>  scala -version
Scala code runner version 2.12.6 -- Copyright 2002-2018,
LAMP/EPFL and Lightbend, Inc.
```

To install Kafka on your machine, ensure that you have at least 4 GB of RAM, and the installation directory will be `/usr/local/kafka/` for macOS users and `/opt/kafka/` for Linux users. Create these directories according to your operating system.

Kafka installation on Linux

Open the Apache Kafka download page, `http://kafka.apache.org/downloads`, as in *Figure 1.1*:

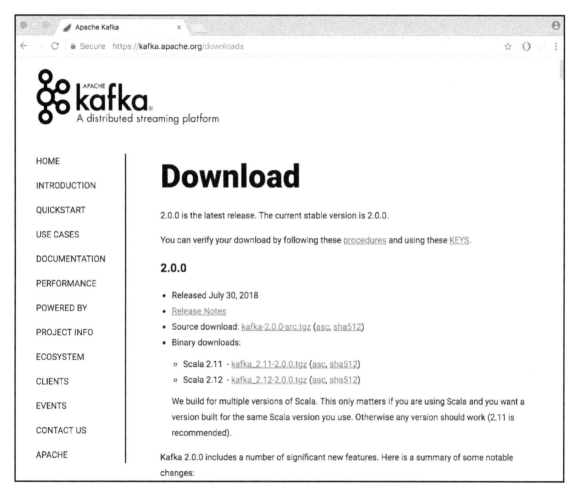

Figure 1.1: Apache Kafka download page

At the time of writing, the current Apache Kafka version is 2.0.0 as a stable release. Remember that, since version 0.8.x, Kafka is not backward-compatible. So, we cannot replace this version for one prior to 0.8. Once you've downloaded the latest available release, let's proceed with the installation.

Remember for macOS users, replace the directory /opt/ with /usr/local.

Follow these steps to install Kafka in Linux:

1. Extract the downloaded file, kafka_2.11-2.0.0.tgz, in the /opt/ directory as follows:

   ```
   > tar xzf kafka_2.11-2.0.0.tgz
   ```

2. Create the KAFKA_HOME environment variable as follows:

   ```
   > export KAFKA_HOME=/opt/kafka_2.11-2.0.0
   ```

3. Add the Kafka bin directory to the PATH variable as follows:

   ```
   > export PATH=$PATH:$KAFKA_HOME/bin
   ```

   Now Java, Scala, and Kafka are installed.

To do all of the previous steps from the command line, there is a powerful tool for macOS users called brew (the equivalent in Linux would be yum).

Kafka installation on macOS

To install from the command line in macOS (brew must be installed), perform the following steps:

1. To install sbt (the Scala build tool) with brew, execute the following:

   ```
   > brew install sbt
   ```

   If already have it in your environment (downloaded previously), run the following to upgrade it:

   ```
   > brew upgrade sbt
   ```

The output is similar to that shown in *Figure 1.2*:

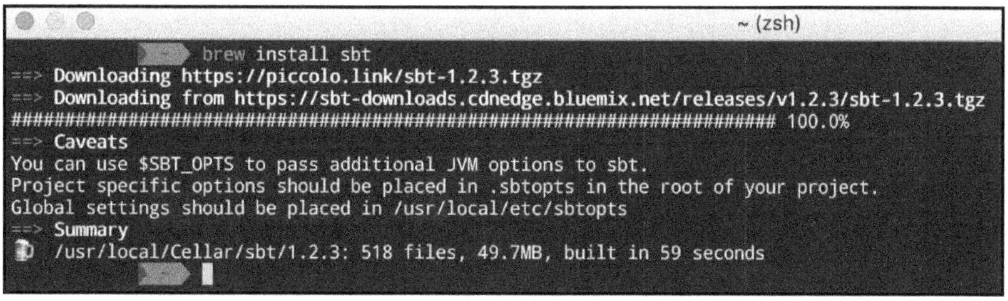

Figure 1.2: The Scala build tool installation output

2. To install Scala with `brew`, execute the following:

> **brew install scala**

If you already have it in your environment (downloaded previously), to upgrade it, run the following command:

> **brew upgrade scala**

The output is similar to that shown in *Figure 1.3*:

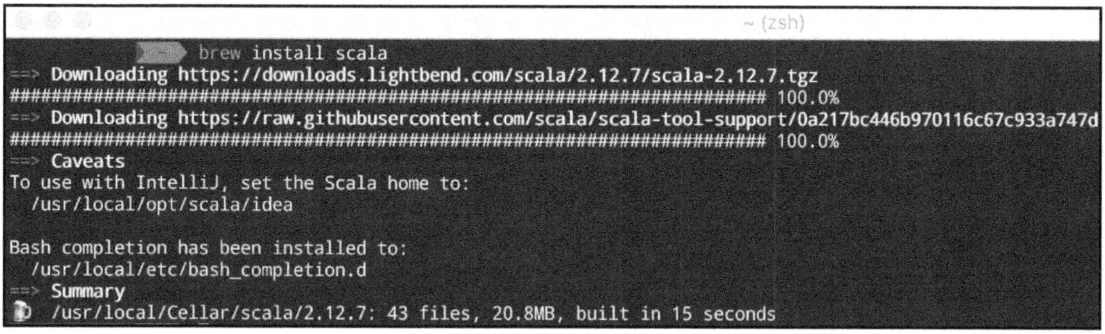

Figure 1.3: The Scala installation output

3. To install Kafka with `brew`, (it also installs Zookeeper), do the following:

> **brew install kafka**

If you already have it (downloaded in the past), upgrade it as follows:

```
> brew upgrade kafka
```

The output is similar to that shown in *Figure 1.4*:

```
                                                                    ~ (zsh)
                    brew install kafka
==> Installing dependencies for kafka: zookeeper
==> Installing kafka dependency: zookeeper
==> Downloading https://homebrew.bintray.com/bottles/zookeeper-3.4.12.high_sierra.bottle.tar.gz
######################################################################## 100.0%
==> Pouring zookeeper-3.4.12.high_sierra.bottle.tar.gz
==> Caveats
To have launchd start zookeeper now and restart at login:
  brew services start zookeeper
Or, if you don't want/need a background service you can just run:
  zkServer start
==> Summary
🍺  /usr/local/Cellar/zookeeper/3.4.12: 242 files, 32.9MB
==> Installing kafka
==> Downloading https://homebrew.bintray.com/bottles/kafka-2.0.0.high_sierra.bottle.tar.gz
######################################################################## 100.0%
==> Pouring kafka-2.0.0.high_sierra.bottle.tar.gz
==> Caveats
To have launchd start kafka now and restart at login:
  brew services start kafka
Or, if you don't want/need a background service you can just run:
  zookeeper-server-start /usr/local/etc/kafka/zookeeper.properties & kafka-server-start /usr/local/etc/kafka/server.properties
==> Summary
🍺  /usr/local/Cellar/kafka/2.0.0: 160 files, 46.8MB
==> Caveats
==> zookeeper
To have launchd start zookeeper now and restart at login:
  brew services start zookeeper
Or, if you don't want/need a background service you can just run:
  zkServer start
==> kafka
To have launchd start kafka now and restart at login:
  brew services start kafka
Or, if you don't want/need a background service you can just run:
  zookeeper-server-start /usr/local/etc/kafka/zookeeper.properties & kafka-server-start /usr/local/etc/kafka/server.properties
```

Figure 1.4: Kafka installation output

Visit `https://brew.sh/` for more about `brew`.

Confluent Platform installation

The third way to install Kafka is through Confluent Platform. In the rest of this book, we will be using Confluent Platform open source version.

Confluent Platform is an integrated platform that includes the following components:

- Apache Kafka
- REST proxy
- Kafka Connect API

- Schema Registry
- Kafka Streams API
- Pre-built connectors
- Non-Java clients
- KSQL

If the reader notices, almost every one of the components has its own chapter in this book.

The commercially licensed Confluent Platform includes, in addition to all of the components of the open source version, the following:

- **Confluent Control Center** (**CCC**)
- Kafka operator (for Kubernetes)
- JMS client
- Replicator
- MQTT proxy
- Auto data balancer
- Security features

It is important to mention that the training on the components of the non-open source version is beyond the scope of this book.

Confluent Platform is available also in Docker images, but here we are going to install it in local.

Open Confluent Platform download page: `https://www.confluent.io/download/` .

At the time of this writing, the current version of Confluent Platform is 5.0.0 as a stable release. Remember that, since the Kafka core runs on Scala, there are two versions: for Scala 2.11 and Scala 2.12.

We could run Confluent Platform from our desktop directory, but following this book's conventions, let's use `/opt/` for Linux users and `/usr/local` for macOS users.

To install Confluent Platform, extract the downloaded file, `confluent-5.0.0-2.11.tar.gz`, in the directory, as follows:

```
> tar xzf confluent-5.0.0-2.11.tar.gz
```

Running Kafka

There are two ways to run Kafka, depending on whether we install it directly or through Confluent Platform.

If we install it directly, the steps to run Kafka are as follows.

 For macOS users, your paths might be different if you've installed using `brew`. Check the output of `brew install kafka` command for the exact command that you can use to start Zookeeper and Kafka.

Go to the Kafka installation directory (`/usr/local/kafka` for macOS users and `/opt/kafka/` for Linux users), as in the example:

```
> cd /usr/local/kafka
```

First of all, we need to start Zookeeper (the Kafka dependency with Zookeeper is and will remain strong). Type the following:

```
> ./bin/zookeeper-server-start.sh ../config/zookeper.properties

ZooKeeper JMX enabled by default

Using config: /usr/local/etc/zookeeper/zoo.cfg

Starting zookeeper ... STARTED
```

To check whether Zookeeper is running, use the `lsof` command over the `9093` port (default port) as follows:

```
> lsof -i :9093

COMMAND PID USER FD TYPE DEVICE SIZE/OFF NODE NAME

java 12529 admin 406u IPv6 0xc41a24baa4fedb11 0t0 TCP *:9093 (LISTEN)
```

Now run the Kafka server that comes with the installation by going to `/usr/local/kafka/` for macOS users and `/opt/kafka/` for Linux users:

```
> ./bin/kafka-server-start.sh ./config/server.properties
```

Now there is an Apache Kafka broker running in your machine.

Remember that Zookeeper must be running on the machine before starting Kafka. If you don't want to start Zookeeper manually every time you need to run Kafka, install it as an operation system auto-start service.

Running Confluent Platform

Go to the Confluent Platform installation directory (`/usr/local/kafka/` for macOS users and `/opt/kafka/` for Linux users) and type the following:

```
> cd /usr/local/confluent-5.0.0
```

To start Confluent Platform, run the following:

```
> bin/confluent start
```

This command-line interface is intended for development only, not for production:

`https://docs.confluent.io/current/cli/index.html`

The output is similar to what is shown in the following code snippet:

```
Using CONFLUENT_CURRENT:
/var/folders/nc/4jrpd1w5563crr_np997zp980000gn/T/confluent.q3uxpyAt

Starting zookeeper
zookeeper is [UP]
Starting kafka
kafka is [UP]
Starting schema-registry
schema-registry is [UP]
Starting kafka-rest
kafka-rest is [UP]
Starting connect
connect is [UP]
Starting ksql-server
ksql-server is [UP]
Starting control-center
control-center is [UP]
```

As indicated by the command output, Confluent Platform automatically starts in this order: Zookeeper, Kafka, Schema Registry, REST proxy, Kafka Connect, KSQL, and the Confluent Control Center.

To access the Confluent Control Center running in your local, go to
`http://localhost:9021`, as shown in *Figure 1.5*:

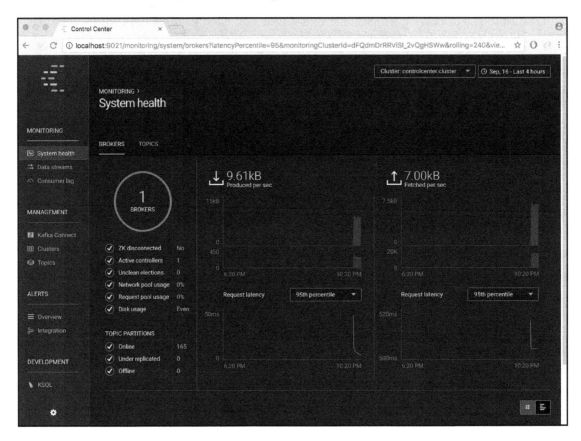

Figure 1.5: Confluent Control Center main page

There are other commands for Confluent Platform.

To get the status of all services or the status of a specific service along with its
dependencies, enter the following:

```
> bin/confluent status
```

To stop all services or a specific service along with the services depending on it, enter the
following:

```
> bin/confluent stop
```

To delete the data and logs of the current Confluent Platform, type the following:

```
> bin/confluent destroy
```

Running Kafka brokers

The real art behind a server is in its configuration. In this section, we will examine how to deal with the basic configuration of a Kafka broker in standalone mode. Since we are learning, at the moment, we will not review the cluster configuration.

As we can suppose, there are two types of configuration: standalone and cluster. The real power of Kafka is unlocked when running with replication in cluster mode and all topics are correctly partitioned.

The cluster mode has two main advantages: parallelism and redundancy. Parallelism is the capacity to run tasks simultaneously among the cluster members. The redundancy warrants that, when a Kafka node goes down, the cluster is safe and accessible from the other running nodes.

This section shows how to configure a cluster with several nodes on our local machine although, in practice, it is always better to have several machines with multiple nodes sharing clusters.

Go to the Confluent Platform installation directory, referenced from now on as <confluent-path>.

As mentioned in the beginning of this chapter, a broker is a server instance. A server (or broker) is actually a process running in the operating system and starts based on its configuration file.

The people of Confluent have kindly provided us with a template of a standard broker configuration. This file, which is called server.properties, is located in the Kafka installation directory in the config subdirectory:

1. Inside <confluent-path>, make a directory with the name mark.

2. For each Kafka broker (server) that we want to run, we need to make a copy of the configuration file template and rename it accordingly. In this example, our cluster is going to be called `mark`:

```
> cp config/server.properties <confluent-
path>/mark/mark-1.properties

> cp config/server.properties <confluent-
path>/mark/mark-2.properties
```

3. Modify each properties file accordingly. If the file is called `mark-1`, the `broker.id` should be `1`. Then, specify the port in which the server will run; the recommendation is `9093` for `mark-1` and `9094` for `mark-2`. Note that the port property is not set in the template, so add the line. Finally, specify the location of the Kafka logs (a Kafka log is a specific archive to store all of the Kafka broker operations); in this case, we use the `/tmp` directory. Here, it is common to have problems with write permissions. Do not forget to give write and execute permissions to the user with whom these processes are executed over the log directory, as in the examples:

- In `mark-1.properties`, set the following:

```
broker.id=1
port=9093
log.dirs=/tmp/mark-1-logs
```

- In `mark-2.properties`, set the following:

```
broker.id=2
port=9094
log.dirs=/tmp/mark-2-logs
```

4. Start the Kafka brokers using the `kafka-server-start` command with the corresponding configuration file passed as the parameter. Don't forget that Confluent Platform must be already running and the ports should not be in use by another process. Start the Kafka brokers as follows:

```
> <confluent-path>/bin/kafka-server-start <confluent-
path>/mark/mark-1.properties &
```

And, in another command-line window, run the following command:

```
> <confluent-path>/bin/kafka-server-start <confluent-
path>/mark/mark-2.properties &
```

Don't forget that the trailing `&` is to specify that you want your command line back. If you want to see the broker output, it is recommended to run each command separately in its own command-line window.

Remember that the properties file contains the server configuration and that the `server.properties` file located in the `config` directory is just a template.

Now there are two brokers, `mark-1` and `mark-2` , running in the same machine in the same cluster.

Remember, there are no dumb questions, as in the following examples:

Q: How does each broker know which cluster it belongs to?

A: The brokers know that they belong to the same cluster because, in the configuration, both point to the same Zookeeper cluster.

Q: How does each broker differ from the others within the same cluster?

A: Every broker is identified inside the cluster by the name specified in the `broker.id` property.

Q: What happens if the port number is not specified?

A: If the port property is not specified, Zookeeper will assign the same port number and will overwrite the data.

Q: What happens if the log directory is not specified?

A: If `log.dir` is not specified, all the brokers will write to the same default `log.dir`. If the brokers are planned to run in different machines, then the port and `log.dir` properties might not be specified (because they run in the same port and log file but in different machines).

Q: How can I check that there is not a process already running in the port where I want to start my broker?

A: As shown in the previous section, there is a useful command to see what process is running on specific port, in this case the `9093` port:

```
> lsof -i :9093
```

The output of the previous command is something like this:

```
COMMAND PID USER FD TYPE DEVICE SIZE/OFF NODE NAME

java 12529 admin 406u IPv6 0xc41a24baa4fedb11 0t0 TCP *:9093 (LISTEN)
```

Your turn: try to run this command before starting the Kafka brokers, and run it after starting them to see the change. Also, try to start a broker on a port in use to see how it fails.

OK, what if I want my cluster to run on several machines?

To run Kafka nodes on different machines but in the same cluster, adjust the Zookeeper connection string in the configuration file; its default value is as follows:

```
zookeeper.connect=localhost:2181
```

Remember that the machines must be able to be found by each other by DNS and that there are no network security restrictions between them.

The default value for Zookeeper connect is correct only if you are running the Kafka broker in the same machine as Zookeeper. Depending on the architecture, it will be necessary to decide if there will be a broker running on the same Zookeeper machine.

To specify that Zookeeper might run in other machines, do the following:

```
zookeeper.connect=localhost:2181, 192.168.0.2:2183, 192.168.0.3:2182
```

The previous line specifies that Zookeeper is running in the local host machine on port 2181, in the machine with IP address 192.168.0.2 on port 2183, and in the machine with IP address, the 192.168.0.3, on port 2182. The Zookeeper default port is 2181, so normally it runs there.

Your turn: as an exercise, try to start a broker with incorrect information about the Zookeeper cluster. Also, using the lsof command, try to raise Zookeeper on a port in use.

If you have doubts about the configuration, or it is not clear what values to change, the server.properties template (as all of the Kafka project) is open sourced in the following:

```
https://github.com/apache/kafka/blob/trunk/config/server.properties
```

Running Kafka topics

The power inside a broker is the topic, namely the queues inside it. Now that we have two brokers running, let's create a Kafka topic on them.

Kafka, like almost all modern infrastructure projects, has three ways of building things: through the command line, through programming, and through a web console (in this case the Confluent Control Center). The management (creation, modification, and destruction) of Kafka brokers can be done through programs written in most modern programming languages. If the language is not supported, it could be managed through the Kafka REST API. The previous section showed how to build a broker using the command line. In later chapters, we will see how to do this process through programming.

Is it possible to only manage (create, modify, or destroy) brokers through programming? No, we can also manage the topics. The topics can also be created through the command line. Kafka has pre-built utilities to manage brokers as we already saw and to manage topics, as we will see next.

To create a topic called `amazingTopic` in our running cluster, use the following command:

```
> <confluent-path>/bin/kafka-topics --create --zookeeper localhost:2181 --replication-factor 1 --partitions 1 --topic amazingTopic
```

The output should be as follows:

```
Created topic amazingTopic
```

Here, the `kafka-topics` command is used. With the `--create` parameter it is specified that we want to create a new topic. The `--topic` parameter sets the name of the topic, in this case, `amazingTopic`.

Do you remember the terms parallelism and redundancy? Well, the `--partitions` parameter controls the parallelism and the `--replication-factor` parameter controls the redundancy.

The `--replication-factor` parameter is fundamental as it specifies in how many servers of the cluster the topic is going to replicate (for example, running). On the other hand, one broker can run just one replica.

Obviously, if a greater number than the number of running servers on the cluster is specified, it will result in an error (you don't believe me? Try it in your environment). The error will be like this:

```
Error while executing topic command: replication factor: 3 larger than
available brokers: 2

[2018-09-01 07:13:31,350] ERROR
org.apache.kafka.common.errors.InvalidReplicationFactorException:
replication factor: 3 larger than available brokers: 2

(kafka.admin.TopicCommand$)
```

To be considered, the broker should be running (don't be shy and test all this theory in your environment).

The `--partitions` parameter, as its name implies, says how many partitions the topic will have. The number determines the parallelism that can be achieved on the consumer's side. This parameter is very important when doing cluster fine-tuning.

Finally, as expected, the `--zookeeper` parameter indicates where the Zookeeper cluster is running.

When a topic is created, the output in the broker log is something like this:

```
[2018-09-01 07:05:53,910] INFO [ReplicaFetcherManager on broker 1] Removed
fetcher for partitions amazingTopic-0 (kafka.server.ReplicaFetcherManager)

[2018-09-01 07:05:53,950] INFO Completed load of log amazingTopic-0 with 1
log segments and log end offset 0 in 21 ms (kafka.log.Log)
```

In short, this message reads like a new topic has been born in our cluster.

How can I check my new and shiny topic? By using the same command: `kafka-topics`.

There are more parameters than `--create`. To check the status of a topic, run the `kafka-topics` command with the `--list` parameter, as follows:

```
> <confluent-path>/bin/kafka-topics.sh --list --zookeeper localhost:2181
```

The output is the list of topics, as we know, is as follows:

```
amazingTopic
```

This command returns the list with the names of all of the running topics in the cluster.

How can I get details of a topic? Using the same command: `kafka-topics`.

For a particular topic, run the `kafka-topics` command with the `--describe` parameter, as follows:

```
> <confluent-path>/bin/kafka-topics --describe --zookeeper localhost:2181 --topic amazingTopic
```

The command output is as follows:

```
Topic:amazingTopic PartitionCount:1 ReplicationFactor:1 Configs: Topic:
amazingTopic Partition: 0 Leader: 1 Replicas: 1 Isr: 1
```

Here is a brief explanation of the output:

- `PartitionCount`: Number of partitions on the topic (parallelism)
- `ReplicationFactor`: Number of replicas on the topic (redundancy)
- `Leader`: Node responsible for reading and writing operations of a given partition
- `Replicas`: List of brokers replicating this topic data; some of these might even be dead
- `Isr`: List of nodes that are currently in-sync replicas

Let's create a topic with multiple replicas (for example, we will run with more brokers in the cluster); we type the following:

```
> <confluent-path>/bin/kafka-topics --create --zookeeper localhost:2181 --replication-factor 2 --partitions 1 --topic redundantTopic
```

The output is as follows:

```
Created topic redundantTopic
```

Now, call the `kafka-topics` command with the `--describe` parameter to check the topic details, as follows:

```
> <confluent-path>/bin/kafka-topics --describe --zookeeper localhost:2181 --topic redundantTopic


Topic:redundantTopic PartitionCount:1 ReplicationFactor:2 Configs:

Topic: redundantTopic Partition: 0 Leader: 1 Replicas: 1,2 Isr: 1,2
```

As you can see, `Replicas` and `Isr` are the same lists; we infer that all of the nodes are in-sync.

Your turn: play with the `kafka-topics` command, and try to create replicated topics on dead brokers and see the output. Also, create topics on running servers and then kill them to see the results. Was the output what you expected?

As mentioned before, all of these commands executed through the command line can be executed programmatically or performed through the Confluent Control Center web console.

A command-line message producer

Kafka also has a command to send messages through the command line; the input can be a text file or the console standard input. Each line typed in the input is sent as a single message to the cluster.

For this section, the execution of the previous steps is needed. The Kafka brokers must be up and running and a topic created inside them.

In a new command-line window, run the following command, followed by the lines to be sent as messages to the server:

```
> <confluent-path>/bin/kafka-console-producer --broker-list localhost:9093
--topic amazingTopic

Fool me once shame on you
Fool me twice shame on me
```

These lines push two messages into the `amazingTopic` running on the localhost cluster on the `9093` port.

This command is also the simplest way to check whether a broker with a specific topic is up and running as it is expected.

As we can see, the `kafka-console-producer` command receives the following parameters:

- `--broker-list`: This specifies the Zookeeper servers specified as a comma-separated list in the form, hostname:port.
- `--topic`: This parameter is followed by the name of the target topic.
- `--sync`: This specifies whether the messages should be sent synchronously.
- `--compression-codec`: This specifies the compression codec used to produce the messages. The possible options are: `none`, `gzip`, `snappy`, or lz4. If not specified, the default is gzip.

- `--batch-size`: If the messages are not sent synchronously, but the message size is sent in a single batch, this value is specified in bytes.
- `--message-send-max-retries`: As the brokers can fail receiving messages, this parameter specifies the number of retries before a producer gives up and drops the message. This number must be a positive integer.
- `--retry-backoff-ms`: In case of failure, the node leader election might take some time. This parameter is the time to wait before producer retries after this election. The number is the time in milliseconds.
- `--timeout`: If the producer is running in asynchronous mode and this parameter is set, it indicates the maximum amount of time a message will queue awaiting for the sufficient batch size. This value is expressed in milliseconds.
- `--queue-size`: If the producer is running in asynchronous mode and this parameter is set, it gives the maximum amount of messages will queue awaiting the sufficient batch size.

In case of a server fine tuning, `batch-size`, `message-send-max-retries`, and `retry-backoff-ms` are very important; take in consideration these parameters to achieve the desired behavior.

If you don't want to type the messages, the command could receive a file where each line is considered a message, as shown in the following example:

```
<confluent-path>/bin/kafka-console-producer --broker-list localhost:9093
-topic amazingTopic < aLotOfWordsToTell.txt
```

A command-line message consumer

The last step is how to read the generated messages. Kafka also has a powerful command that enables messages to be consumed from the command line. Remember that all of these command-line tasks can also be done programmatically. As the producer, each line in the input is considered a message from the producer.

For this section, the execution of the previous steps is needed. The Kafka brokers must be up and running and a topic created inside them. Also, some messages need to be produced with the message console producer, to begin consuming these messages from the console.

Run the following command:

```
> <confluent-path>/bin/kafka-console-consumer --topic amazingTopic --
bootstrap-server localhost:9093 --from-beginning
```

The output should be as follows:

```
Fool me once shame on you
Fool me twice shame on me
```

The parameters are the topic's name and the name of the broker producer. Also, the --from-beginning parameter indicates that messages should be consumed from the beginning instead of the last messages in the log (now test it, generate many more messages, and don't specify this parameter).

There are more useful parameters for this command, some important ones are as follows:

- --fetch-size: This is the amount of data to be fetched in a single request. The size in bytes follows as argument. The default value is 1,024 x 1,024.
- --socket-buffer-size: This is the size of the TCP RECV. The size in bytes follows this parameter. The default value is 2 x 1024 x 1024.
- --formater: This is the name of the class to use for formatting messages for display. The default value is NewlineMessageFormatter.
- --autocommit.interval.ms: This is the time interval at which to save the current offset in milliseconds. The time in milliseconds follows as argument. The default value is 10,000.
- --max-messages: This is the maximum number of messages to consume before exiting. If not set, the consumption is continuous. The number of messages follows as the argument.
- --skip-message-on-error: If there is an error while processing a message, the system should skip it instead of halting.

The most requested forms of this command are as follows:

- To consume just one message, use the following:

```
> <confluent-path>/bin/kafka-console-consumer --topic
amazingTopic --
bootstrap-server localhost:9093 --max-messages 1
```

- To consume one message from an offset, use the following:

```
> <confluent-path>/bin/kafka-console-consumer --topic
amazingTopic --
bootstrap-server localhost:9093 --max-messages 1 --formatter
'kafka.coordinator.GroupMetadataManager$OffsetsMessageFormatter'
```

- To consume messages from a specific consumer group, use the following:

```
<confluent-path>/bin/kafka-console-consumer -topic amazingTopic -
- bootstrap-server localhost:9093 --new-consumer --consumer-
property
group.id=my-group
```

Using kafkacat

kafkacat is a generic command-line non-JVM utility used to test and debug Apache Kafka deployments. kafkacat can be used to produce, consume, and list topic and partition information for Kafka. kafkacat is netcat for Kafka, and it is a tool for inspecting and creating data in Kafka.

kafkacat is similar to the Kafka console producer and Kafka console consumer, but more powerful.

kafkacat is an open source utility and it is not included in Confluent Platform. It is available at https://github.com/edenhill/kafkacat.

To install kafkacat on modern Linux, type the following:

```
apt-get install kafkacat
```

To install kafkacat on macOS with brew, type the following:

```
brew install kafkacat
```

To subscribe to amazingTopic and redundantTopic and print to stdout, type the following:

```
kafkacat -b localhost:9093 -t amazingTopic redundantTopic
```

Summary

In this chapter, we've learned what Kafka is, how to install and run Kafka in Linux and macOS and how to install and run Confluent Platform.

Also, we've reviewed how to run Kafka brokers and topics, how to run a command-line message producer and consumer, and how to use kafkacat.

In `Chapter 2`, *Message Validation*, we will analyze how to build a producer and a consumer from Java.

Message Validation 2

Chapter 1, *Configuring Kafka*, focused on how to set up a Kafka cluster and run a command-line producer and a consumer. Having the event producer, we now have to process those events.

Before going into detail, let's present our case study. We need to model the systems of Monedero, a fictional company whose core business is cryptocurrency exchange. Monedero wants to base its IT infrastructure on an **enterprise service bus** (**ESB**) built with Apache Kafka. The Monedero IT department wants to unify the service backbone across the organization. Monedero also has worldwide, web-based, and mobile-app-based clients, so a real-time response is fundamental.

Online customers worldwide browse the Monedero website to exchange their cryptocurrencies. There are a lot of use cases that customers can perform in Monedero, but this example is focused on the part of the exchange workflow specifically from the web application.

This chapter covers the following topics:

- Modeling the messages in JSON format
- Setting up a Kafka project with Gradle
- Reading from Kafka with a Java client
- Writing to Kafka with a Java client
- Running a processing engine pipeline
- Coding a `Validator` in Java
- Running the validation

Enterprise service bus in a nutshell

Event processing consists of taking one or more events from an event stream and applying actions over those events. In general, in an enterprise service bus, there are commodity services; the most common are the following:

- Data transformation
- Event handling
- Protocol conversion
- Data mapping

Message processing in the majority of cases involves the following:

- Message structure validation against a message schema
- Given an event stream, filtering the messages from the stream
- Message enrichment with additional data
- Message aggregation (composition) from two or more message to produce a new message

This chapter is about event validation. The chapters that follow are about composition and enrichment.

Event modeling

The first step in event modeling is to express the event in English in the following form:

Subject-verb-direct object

For this example, we are modeling the event *customer consults the ETH price*:

- The subject in this sentence is *customer*, a noun in nominative case. The subject is the entity performing the action.
- The verb in this sentence is *consults*; it describes the action performed by the subject.
- The direct object in this sentence is *ETH price*. The object is the entity in which the action is being done.

We can represent our message in several message formats (covered in other sections of this book):

- **JavaScript Object Notation (JSON)**
- Apache Avro
- Apache Thrift
- Protocol Buffers

JSON is easily read and written by both humans and machines. For example, we could chose binary as the representation, but it has a rigid format and it was not designed for humans to read it; as counterweight, binary representation is very fast and lightweight in processing.

Listing 2.1, shows the representation of the CUSTOMER_CONSULTS_ETHPRICE event in JSON format:

```
{
    "event": "CUSTOMER_CONSULTS_ETHPRICE",
    "customer": {
            "id": "14862768",
            "name": "Snowden, Edward",
            "ipAddress": "95.31.18.111"
    },
    "currency": {
            "name": "ethereum",
            "price": "RUB"
    },
    "timestamp": "2018-09-28T09:09:09Z"
}
```

Listing 2.1: customer_consults_ETHprice.json

For this example, the **Ethereum (ETH)** currency price is expressed in **Russian rouble (RUB)**. This JSON message has four sections:

- event: This is a string with the event's name.
- customer: This represents the person (in this case its id is 14862768) consulting the Ethereum price. In this representation, there is a unique id for the customer, the name, and the browser ipAddress, which is the IP address of the computer the customer is logged on.
- currency: This contains the cryptocurrency name and the currency in which the price is expressed.
- timestamp: The timestamp in which the customer made the request (UTC).

From another perspective, the message has two parts: the metadata—this is the event name and the timestamp and two business entities, the customer and the currency. As we can see, this message can be read and understood by a human.

Other messages from the same use case in JSON format could be as follows:

```
{ "event": "CUSTOMER_CONSULTS_ETHPRICE",
  "customer": {
        "id": "13548310",
        "name": "Assange, Julian",
        "ipAddress": "185.86.151.11"
  },
  "currency": {
        "name": "ethereum",
        "price": "EUR"
  },
  "timestamp": "2018-09-28T08:08:14Z"
}
```

This is another example:

```
{ "event": "CUSTOMER_CONSULTS_ETHPRICE",
  "customer": {
        "id": "15887564",
        "name": "Mills, Lindsay",
        "ipAddress": "186.46.129.15"
  },
  "currency": {
        "name": "ethereum",
        "price": "USD"
  },
  "timestamp": "2018-09-28T19:51:35Z"
}
```

What happens if we want to represent our message in the Avro schema? Yes, the Avro schema of our message (note that it's not the message, but the schema) is in *Listing 2.2*:

```
{ "name": "customer_consults_ethprice",
  "namespace": "monedero.avro",
  "type": "record",
  "fields": [
    { "name": "event", "type": "string" },
    { "name": "customer",
      "type": {
          "name": "id", "type": "long",
          "name": "name", "type": "string",
          "name": "ipAddress", "type": "string"
      }
```

```
    },
    { "name": "currency",
      "type": {
          "name": "name", "type": "string",
          "name": "price", "type": {
          "type": "enum", "namespace": "monedero.avro",
            "name": "priceEnum", "symbols": ["USD", "EUR", "RUB"]}
      }
    },
    { "name": "timestamp", "type": "long",
      "logicalType": "timestamp-millis"
    }
  ]
}
```

Listing 2.2: customer_consults_ethprice.avsc

For more information about the Avro schema, check the Apache Avro specification:

`https://avro.apache.org/docs/1.8.2/spec.html`

Setting up the project

This time, we are going to build our project with Gradle. The first step is to download and install Gradle from `http://www.gradle.org/downloads`.

Gradle only requires a Java JDK (version 7 or higher).

macOS users can install Gradle with the `brew` command, as follows:

```
$ brew update && brew install gradle
```

The output is something like the following:

```
==> Downloading
https://services.gradle.org/distributions/gradle-4.10.2-all.zip
==> Downloading from
https://downloads.gradle.org/distributions/gradle-4.10.2-al
######################################################################
100.0%
  /usr/local/Cellar/gradle/4.10.2: 203 files, 83.7MB, built in 59 seconds
```

Linux users can install Gradle with the `apt-get` command, as follows:

```
$ apt-get install gradle
```

Unix users can install with sdkman, a tool for managing parallel versions of most Unix-based systems, as follows:

```
$ sdk install gradle 4.3
```

To check that Gradle is installed correctly, type the following:

```
$ gradle -v
```

The output is something like the following:

```
------------------------------------------------------------
Gradle 4.10.2
------------------------------------------------------------
```

The first step is to create a directory called `monedero` and, from that directory, execute the following:

```
$ gradle init --type java-library
```

The output is something like the following:

```
. . .
BUILD SUCCESSFUL
. . .
```

Gradle generates a skeleton project inside the directory. The directory should be similar to the following:

- – `build.gradle`
- – `gradle`
- –– `wrapper`
- ––– `gradle-wrapper.jar`
- ––– `gradle-vreapper.properties`
- – `gradlew`
- – `gradle.bat`
- – `settings.gradle`
- – `src`

- -- main
- --- java
- ----- Library.java
- -- test
- --- java
- ----- LibraryTest.java

The two Java files, `Library.java` and `LibraryTest.java`, can be deleted.

Now, modify the Gradle build file called `build.gradle`, and replace it with *Listing 2.3*:

```
apply plugin: 'java'
apply plugin: 'application'
sourceCompatibility = '1.8'
mainClassName = 'monedero.ProcessingEngine'
repositories {
 mavenCentral()
}
version = '0.1.0'
dependencies {
  compile group: 'org.apache.kafka', name: 'kafka_2.12', version: '2.0.0'
  compile group: 'com.fasterxml.jackson.core', name: 'jackson-core',
version: '2.9.7'
}
jar {
 manifest {
 attributes 'Main-Class': mainClassName
 } from {
 configurations.compile.collect {
 it.isDirectory() ? it : zipTree(it)
 }
 }
 exclude "META-INF/*.SF"
 exclude "META-INF/*.DSA"
 exclude "META-INF/*.RSA"
}
```

Listing 2.3: ProcessingEngine Gradle build file

This file shows the library dependencies for the engine:

- `kafka_2.12`, are the dependencies for Apache Kafka
- `jackson-databind` is the library for JSON parsing and manipulation

To compile the sources and download the required libraries, type the command:

```
$ gradle compileJava
```

The output is something like the following:

```
. . .
BUILD SUCCESSFUL
. . .
```

The project can be created with Maven or SBT, even from an IDE (IntelliJ, Eclipse, Netbeans). But for simplicity here, it was created with Gradle.

For more information about the build tools, visit the following links:

- Gradle's main page: http://www.gradle.org
- Maven's main page: http://maven.apache.org
- SBT's main page: http://www.scala-sbt.org/

Reading from Kafka

Now that we have our project skeleton, let's recall the project requirements for the stream processing engine. Remember that our event customer consults ETH price occurs outside Monedero and that these messages may not be well formed, that is, they may have defects. The first step in our pipeline is to validate that the input events have the correct data and the correct structure. Our project will be called `ProcessingEngine`.

The `ProcessingEngine` specification shall create a pipeline application that does the following:

- Reads each message from a Kafka topic called **input-messages**
- Validates each message, sending any invalid event to a specific Kafka topic called **invalid-messages**
- Writes the correct messages in a Kafka topic called **valid-messages**

These steps are detailed in *Figure 2.1*, the first sketch for the pipeline processing engine:

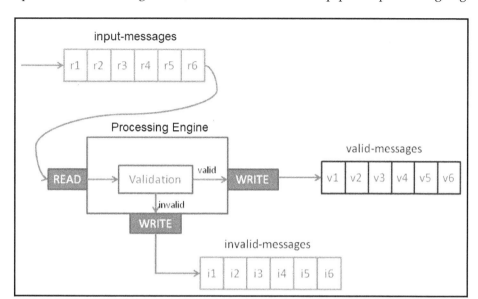

Figure 2.1: The processing engine reads events from the input-messages topic, validates the messages, and routes the defective ones to invalid-messages topic and the correct ones to valid-messages topic

The processing engine stream construction has two phases:

- Create a simple Kafka worker that reads from the **input-messages** topic in Kafka and writes the events to another topic
- Modify the Kafka worker to make the validation

So, let's proceed with the first step. Build a Kafka worker that reads individual raw messages from the **input-messages** topic. We say in the Kafka jargon that a consumer is needed. If you recall, in the first chapter we built a command-line producer to write events to a topic and a command-line consumer to read the events from that topic. Now, we will code the same consumer in Java.

For our project, a consumer is a Java interface that contains all of the necessary behavior for all classes that implement consumers.

Create a file called `Consumer.java` in the `src/main/java/monedero/` directory with the content of *Listing 2.4*:

```
package monedero;
import java.util.Properties;
public interface Consumer {
  static Properties createConfig(String servers, String groupId) {
    Properties config = new Properties();
    config.put("bootstrap.servers", servers);
    config.put("group.id", groupId);
    config.put("enable.auto.commit", "true");
    config.put("auto.commit.interval.ms", "1000");
    config.put("auto.offset.reset", "earliest");
    config.put("session.timeout.ms", "30000");
    config.put("key.deserializer",
        "org.apache.kafka.common.serialization.StringDeserializer");
    config.put("value.deserializer",
        "org.apache.kafka.common.serialization.StringDeserializer");
    return config;
  }
}
```

Listing 2.4: Consumer.java

The consumer interface encapsulates the common behavior of the Kafka consumers. The consumer interface has the `createConfig` method that sets all of the properties needed by all of the Kafka consumers. Note that the deserializers are of the `StringDeserializer` type because the Kafka consumer reads Kafka key-value records where the value are of the type string.

Now, create a file called `Reader.java` in the `src/main/java/monedero/` directory with the content of *Listing 2.5*:

```
package monedero;
import org.apache.kafka.clients.consumer.ConsumerRecord;
import org.apache.kafka.clients.consumer.ConsumerRecords;
import org.apache.kafka.clients.consumer.KafkaConsumer;
import java.time.Duration;
import java.util.Collections;
class Reader implements Consumer {
  private final KafkaConsumer<String, String> consumer;//1
  private final String topic;
  Reader(String servers, String groupId, String topic) {
    this.consumer =
        new KafkaConsumer<>(Consumer.createConfig(servers, groupId));
    this.topic = topic;
  }
```

```
void run(Producer producer) {
    this.consumer.subscribe(Collections.singletonList(this.topic));//2
    while (true) {//3
        ConsumerRecords<String, String> records =
consumer.poll(Duration.ofMillis(100));   //4
        for (ConsumerRecord<String, String> record : records) {
producer.process(record.value());//5
        }
    }
  }
}
```

Listing 2.5: Reader.java

The `Reader` class implements the consumer interface. So, `Reader` is a Kafka consumer:

- In line `//1`, `<String, String>` says that `KafkaConsumer` reads Kafka records where the key and value are both of the type string
- In line `//2`, the consumer subscribes to the Kafka topic specified in its constructor
- In line `//3`, there is a `while(true)` infinite loop for demonstrative purposes; in practice, we need to deal with more robust code maybe, implementing `Runnable`
- In line `//4`, this consumer will be pooling data from the specified topics every 100 milliseconds
- In line `//5`, the consumer sends the message to be processed by the producer

This consumer reads all of the messages from the specified Kafka topic and sends them to the process method of the specified producer. All of the configuration properties are specified in the consumer interface, but specifically the `groupId` property is important because it associates the consumer with a specific consumer group.

The consumer group is useful when we need to share the topic's events across all of the group's members. Consumer groups are also used to group or isolate different instances.

To read more about the Kafka Consumer API, follow this link: `https://kafka.apache.org/20/javadoc/org/apache/kafka/clients/consumer/KafkaConsumer.html/`

Writing to Kafka

Our `Reader` invokes the `process()` method; this method belonging to the `Producer` class. As with the consumer interface, the producer interface encapsulates all of the common behavior of the Kafka producers. The two producers in this chapter implement this producer interface.

In a file called `Producer.java`, located in the `src/main/java/monedero` directory, copy the content of *Listing 2.6*:

```
package monedero;
import java.util.Properties;
import org.apache.kafka.clients.producer.KafkaProducer;
import org.apache.kafka.clients.producer.ProducerRecord;
public interface Producer {
  void process(String message);                              //1
  static void write(KafkaProducer<String, String> producer,
                    String topic, String message) {          //2
    ProducerRecord<String, String> pr = new ProducerRecord<>(topic,
message);
    producer.send(pr);
  }
  static Properties createConfig(String servers) {           //3
    Properties config = new Properties();
    config.put("bootstrap.servers", servers);
    config.put("acks", "all");
    config.put("retries", 0);
    config.put("batch.size", 1000);
    config.put("linger.ms", 1);
    config.put("key.serializer",
"org.apache.kafka.common.serialization.StringSerializer");
config.put("value.serializer",
        "org.apache.kafka.common.serialization.StringSerializer");
        return config;
  }
}
```

Listing 2.6: Producer.java

The producer interface has the following observations:

- An abstract method called process invoked in the `Reader` class
- A static method called write that sends a message to the producer in the specified topic
- A static method called `createConfig`, where it sets all of the properties required for a generic producer

As with the consumer interface, an implementation of the producer interface is needed. In this first version, we just pass the incoming messages to another topic without modifying the messages. The implementation code is in *Listing 2.7* and should be saved in a file called `Writer.java` in the `src/main/java/m` directory.

The following is the content of *Listing 2.7*, `Writer.java`:

```
package monedero;
import org.apache.kafka.clients.producer.KafkaProducer;
public class Writer implements Producer {
  private final KafkaProducer<String, String> producer;
  private final String topic;
  Writer(String servers, String topic) {
    this.producer = new KafkaProducer<>(
        Producer.createConfig(servers));//1
    this.topic = topic;
  }
  @Override
  public void process(String message) {
    Producer.write(this.producer, this.topic, message);//2
  }
}
```

Listing 2.7: Writer.java

In this implementation of the `Producer` class, we see the following:

- The `createConfig` method is invoked to set the necessary properties from the producer interface
- The process method writes each incoming message in the output topic. As the message arrives from the topic, it is sent to the target topic

This producer implementation is very simple; it doesn't modify, validate, or enrich the messages. It just writes them to the output topic.

To read more about the Kafka producer API, follow this link:

```
https://kafka.apache.org/0110/javadoc/index.html?org/apache/kafka/clients/
consumer/KafkaProducer.html
```

Running the processing engine

The `ProcessingEngine` class coordinates the `Reader` and `Writer` classes. It contains the main method to coordinate them. Create a new file called `ProcessingEngine.java` in the `src/main/java/monedero/` directory and copy therein the code in *Listing 2.8*.

The following is the content of *Listing 2.8*, `ProcessingEngine.java`:

```
package monedero;
public class ProcessingEngine {
  public static void main(String[] args) {
    String servers = args[0];
    String groupId = args[1];
    String sourceTopic = args[2];
    String targetTopic = args[3];
    Reader reader = new Reader(servers, groupId, sourceTopic);
    Writer writer = new Writer(servers, targetTopic);
    reader.run(writer);
  }
}
```

Listing 2.8: ProcessingEngine.java

`ProcessingEngine` receives four arguments from the command line:

- `args[0]` `servers`, the host and port of the Kafka broker
- `args[1]` `groupId`, the consumer group of the consumer
- `args[2]` `sourceTopic`, input Topic where `Reader` reads from
- `args[3]` `targetTopic`, output Topic where `Writer` writes to

To build the project, run this command from the `monedero` directory:

```
$ gradle jar
```

If everything is OK, the output is something like the following:

```
. . .
BUILD SUCCESSFUL
. . .
```

To run the project, we need to open three different command-line windows. *Figure 2.2* shows what the command-line windows should look:

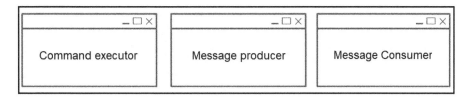

Figure 2.2: The three terminal windows to test the processing engine including message producer, message consumer, and the application itself

1. In the first command-line terminal, move to the `Confluent` directory and start it, as follows:

   ```
   $ bin/confluent start
   ```

2. Once the control center (Zookeeper and Kafka included) is running in the same command-line terminal, generate the two necessary topics, as follows:

   ```
   $ bin/kafka-topics --create --zookeeper localhost:2181 --
   replication-factor 1 --partitions 1 --topic input-topic
   ```

   ```
   $ bin/kafka-topics --create --zookeeper localhost:2181 --
   replication-factor 1 --partitions 1 --topic output-topic
   ```

   Recall, to display the topics running in our cluster type, use the following:

   ```
   $ bin/kafka-topics --list --zookeeper localhost:2181
   ```

   If you had a mistyping error, to delete some topic (just in case), type the following:

   ```
   $ bin/kafka-topics --delete --zookeeper localhost:2181 --topic
   unWantedTopic
   ```

3. In the same command-line terminal, start the console producer running the input-topic topic, as follows:

```
$ bin/kafka-console-producer --broker-list localhost:9092 --topic
input-topic
```

This window is where the input messages are typed.

4. In the second command-line terminal, start a console consumer listening to output-topic by typing the following:

```
$ bin/kafka-console-consumer --bootstrap-server localhost:9092 --
from-beginning --topic output-topic
```

5. In the third command-line terminal, start the processing engine. Go to the project root directory where the gradle jar command was executed and run, as follows:

```
$ java -jar ./build/libs/monedero-0.1.0.jar localhost:9092 foo
input-topic output-topic
```

Now, the show consists in reading all of the events from input-topic and writing them in output-topic.

Go to the first command-line terminal (the message producer) and send the following three messages (remember to type enter between messages and execute each one in just one line):

```
{"event": "CUSTOMER_CONSULTS_ETHPRICE", "customer": {"id": "14862768",
"name": "Snowden, Edward", "ipAddress": "95.31.18.111"}, "currency":
{"name": "ethereum", "price": "RUB"}, "timestamp": "2018-09-28T09:09:09Z"}

{"event": "CUSTOMER_CONSULTS_ETHPRICE", "customer": {"id": "13548310",
"name": "Assange, Julian", "ipAddress": "185.86.151.11"}, "currency":
{"name": "ethereum", "price": "EUR"}, "timestamp": "2018-09-28T08:08:14Z"}

{"event": "CUSTOMER_CONSULTS_ETHPRICE", "customer": {"id": "15887564",
"name": "Mills, Lindsay", "ipAddress": "186.46.129.15"}, "currency":
{"name": "ethereum", "price": "USD"}, "timestamp": "2018-09-28T19:51:35Z"}
```

If everything is working fine, the messages typed in the console-producer should be appearing in the console-consumer window, because the processing engine is copying from input-topic to output-topic.

The next step is to move onto a more complex version involving message validation (this current chapter), message enrichment (Chapter 3, *Message Enrichment*), and message transformation (Chapter 4, *Serialization*).

Using the same suggestion made in the `Chapter 1`, *Configuring Kafka*, the replication-factor and partitions parameters where set to 1; try setting different values and see what happens when you stop one server.

Coding a validator in Java

The `Writer` class implements the producer interface. The idea is to modify that `Writer` and build a validation class with minimum effort. The `Validator` process is as follows:

- Read the Kafka messages from the **input-messages** topic
- Validate the messages, sending defective messages to the **invalid-messages** topic
- Write the well-formed messages to **valid-messages** topic

At the moment, for this example, the definition of a valid message is a message t0 which the following applies:

- It is in JSON format
- It contains the four required fields: event, customer, currency, and timestamp

If these conditions are not met, a new error message in JSON format is generated, sending it to the invalid-messages Kafka topic. The schema of this error message is very simple:

```
{"error": "Failure description" }
```

The first step is create a new `Validator.java` file in the `src/main/java/monedero/` directory, and copy therein the content of *Listing 2.9*.

The following is the content of *Listing 2.9*, `Validator.java`:

```
package monedero;
import com.fasterxml.jackson.databind.JsonNode;
import com.fasterxml.jackson.databind.ObjectMapper;
import org.apache.kafka.clients.producer.KafkaProducer;
import java.io.IOException;

public class Validator implements Producer {
  private final KafkaProducer<String, String> producer;
  private final String validMessages;
  private final String invalidMessages;
  private static final ObjectMapper MAPPER = new ObjectMapper();
  public Validator(String servers, String validMessages, String
invalidMessages) { //1
    this.producer = new KafkaProducer<>(Producer.createConfig(servers));
```

```
      this.validMessages = validMessages;
      this.invalidMessages = invalidMessages;
    }
    @Override
    public void process(String message) {
      try {
        JsonNode root = MAPPER.readTree(message);
        String error = "";
        error = error.concat(validate(root, "event")); //2
        error = error.concat(validate(root, "customer"));
        error = error.concat(validate(root, "currency"));
        error = error.concat(validate(root, "timestamp"));
        if (error.length() > 0) {
          Producer.write(this.producer, this.invalidMessages, //3
          "{\"error\": \" " + error + "\"}");
        } else {
          Producer.write(this.producer, this.validMessages, //4
          MAPPER.writeValueAsString(root));
        }
      } catch (IOException e) {
        Producer.write(this.producer, this.invalidMessages, "{\"error\": \""
        + e.getClass().getSimpleName() + ": " + e.getMessage() + "\"}");//5
      }
    }
  private String validate(JsonNode root, String path) {
    if (!root.has(path)) {
      return path.concat(" is missing. ");
    }
    JsonNode node = root.path(path);
    if (node.isMissingNode()) {
      return path.concat(" is missing. ");
    }
    return "";
  }
}
```

Listing 2.9: Validator.java

As with `Writer`, the `Validator` class also implements the `Producer` class, but with the following:

- In line `//1`, its constructor takes two topics: the valid and the invalid-messages topic
- In line `//2`, the process method validates the fact that the message is in JSON format along with the existence of the fields: event, customer, currency, and timestamp

- In line //3, in case the message doesn't have any required field, an error message is sent to the invalid-messages topic
- In line //4, in case the message is valid, the message is sent to the valid-messages topic
- In line //5, in case the message is not in JSON format, an error message is sent to the invalid-messages topic

Running the validation

At the moment, the `ProcessingEngine` class coordinates the `Reader` and `Writer` classes. It contains the main method to coordinate them. We have to edit the `ProcessingEngine` class located in the `src/main/java/monedero/` directory and change `Writer` with `Validator`, as in *Listing 2.10*.

The following is the content of *Listing 2.10*, `ProcessingEngine.java`:

```
package monedero;
public class ProcessingEngine {
  public static void main(String[] args) {
    String servers = args[0];
    String groupId = args[1];
    String inputTopic = args[2];
    String validTopic = args[3];
    String invalidTopic = args[4];
    Reader reader = new Reader(servers, groupId, inputTopic);
    Validator validator = new Validator(servers, validTopic, invalidTopic);
    reader.run(validator);
  }
}
```

Listing 2.10: ProcessingEngine.java

`ProcessingEngine` receives five arguments from the command line:

- `args[0]` `servers`, indicates the host and port of the Kafka broker
- `args[1]` `groupId`, indicates that the consumer is part of this Kafka consumer group
- `args[2]` `inputTopic`, the topic where `Reader` reads from
- `args[3]` `validTopic`, the topic where valid messages are sent
- `args[4]` `invalidTopic`, the topic where invalid messages are sent

To rebuild the project from the `monedero` directory, run the following command:

```
$ gradle jar
```

If everything is OK, the output should be similar to the following:

```
. . .
BUILD SUCCESSFUL
. . .
```

To run the project, we need four different command-line windows. *Figure 2.3* shows the command-line windows arrangement:

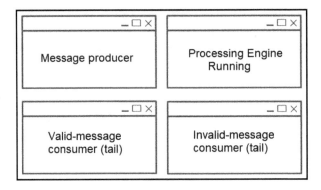

Figure 2.3: The four terminal windows to test the processing engine including: message producer, valid-message consumer, invalid-message consumer, and the processing engine itself

1. In the first command-line terminal, go to the Kafka installation directory and generate the two necessary topics:

```
$ bin/kafka-topics --create --zookeeper localhost:2181 --
replication-factor 1 --partitions 1 --topic valid-messages
```

```
$ bin/kafka-topics --create --zookeeper localhost:2181 --
replication-factor 1 --partitions 1 --topic invalid-messages
```

Then, start a console producer to the `input-topic` topic, as follows:

```
$ bin/kafka-console-producer --broker-list localhost:9092 --topic
input-topic
```

This window is where the input messages are produced (typed).

2. In the second command-line window, start a command-line consumer listening to the valid-messages topic, as follows:

```
$ bin/kafka-console-consumer --bootstrap-server localhost:9092 --
from-beginning --topic valid-messages
```

3. In the third command-line window, start a command-line consumer listening to invalid-messages topic, as follows:

```
$ bin/kafka-console-consumer --bootstrap-server localhost:9092 --
from-beginning --topic invalid-messages
```

4. In the fourth command-line terminal, start up the processing engine. From the project root directory (where the `gradle jar` command were executed), run the following command:

```
$ java -jar ./build/libs/monedero-0.1.0.jar localhost:9092 foo
input-topic valid-messages invalid-messages
```

From the first command-line terminal (the console producer), send the following three messages (remember to type enter between messages and execute each one in just one line):

```
{"event": "CUSTOMER_CONSULTS_ETHPRICE", "customer": {"id": "14862768",
"name": "Snowden, Edward", "ipAddress": "95.31.18.111"}, "currency":
{"name": "ethereum", "price": "RUB"}, "timestamp": "2018-09-28T09:09:09Z"}

{"event": "CUSTOMER_CONSULTS_ETHPRICE", "customer": {"id": "13548310",
"name": "Assange, Julian", "ipAddress": "185.86.151.11"}, "currency":
{"name": "ethereum", "price": "EUR"}, "timestamp": "2018-09-28T08:08:14Z"}

{"event": "CUSTOMER_CONSULTS_ETHPRICE", "customer": {"id": "15887564",
"name": "Mills, Lindsay", "ipAddress": "186.46.129.15"}, "currency":
{"name": "ethereum", "price": "USD"}, "timestamp": "2018-09-28T19:51:35Z"}
```

As these are valid messages, the messages typed in the producer console should appear in the valid-messages consumer console window.

Now try sending defective messages; first, try messages that are not in JSON format:

```
I am not JSON, I am Freedy. [enter]
I am a Kafkeeter! [enter]
```

This message should be received in the invalid messages topic (and displayed in the window), as in the following example:

```
{"error": "JsonParseException: Unrecognized token ' I am not JSON, I am
Freedy.': was expecting 'null','true', 'false' or NaN
at [Source: I am not JSON, I am Freedy.; line: 1, column: 4]"}
```

Then, let's try something more complex, the first message but without a timestamp, as in the example:

```
{"event": "CUSTOMER_CONSULTS_ETHPRICE", "customer": {"id": "14862768",
"name": "Snowden, Edward", "ipAddress": "95.31.18.111"}, "currency":
{"name": "ethereum", "price": "RUB"}}
```

This message should be received in the invalid messages topic, as follows:

```
{"error": "timestamp is missing."}
```

The message validation is complete and, as you can see, there is a lot more validation to do, for example, validation against JSON schemas, but this is covered in `Chapter 5`, *Schema Registry*.

The architecture detailed in *Figure 2.1* of this chapter will be used in `Chapter 3`, *Message Enrichment*.

Summary

In this chapter we learned how to model the messages in JSON format and how to set up a Kafka project with Gradle.

Also, we learned how to write to and read from Kafka with a Java client, how to run the processing engine, how to code a validator in Java, and how to run the message validation.

In `Chapter 3`, *Message Enrichment*, the architecture of this chapter will be redesigned to incorporate message enrichment.

Message Enrichment

To fully understand this chapter, it is necessary to have read the previous chapter that focused on how to validate events. This chapter is focused on how to enrich events.

In this chapter, we will continue using the systems of Monedero, our fictitious company that is dedicated to the exchange of cryptocurrencies. If we remember in the previous chapter, the messages of Monedero were validated; in this chapter, we will continue with the same flow, but we will add one more step of enrichment.

In this context, we understand enrichment as adding extra data that was not in the original message. In this chapter, we will see how to enrich a message with geographic location using the MaxMind database and how to extract the current value of the exchange rate using the Open Exchange data. If we remember the events that we modeled for Monedero, each one included the IP address of the customer's computer.

In this chapter, we will use the MaxMind free database that provides us with an API that contains a mapping of IP addresses to their geographic location.

Our system in Monedero searches for the IP address of our customer in the MaxMind database to determine where the customer is located when the request to our system was made. The use of data from external sources to add them to our events is what we call message enrichment.

In the cryptocurrencies world, there is something called Bit License, in which some geographic areas are limited by law to carry out activities with cryptocurrencies. We currently have an event validation service for Monedero.

However, the legal department has asked us to have a validation filter to know the geographic location of our customers and thus be able to comply with the Bit License. The Bit License has operated in the New York area since July 2014 and applies to residents. Under the terms of the law, those considered resident are all of the people who reside, are located, have a place of business, or conduct business in the state of New York.

This chapter covers the following topics:

- How extraction works
- How enrichment works
- Extracting the location given an IP address
- Extracting the currency price given a currency
- Extracting the weather data given a location
- Enriching messages with the geographic location
- Enriching messages with a currency price
- Running a processing engine

Extracting the geographic location

Open the `build.gradle` file on the Monedero project created in `Chapter 2`, *Message Validation*, and add the lines highlighted in *Listing 3.1*.

The following is the content of *Listing 3.1*, the Monedero `build.gradle` file:

```
apply plugin: 'java'
apply plugin: 'application'
sourceCompatibility = '1.8'
mainClassName = 'monedero.ProcessingEngine'
repositories {
  mavenCentral()
}
version = '0.2.0'
dependencies {
    compile group: 'org.apache.kafka', name: 'kafka_2.12', version:
                                                            '2.0.0'
    compile group: 'com.maxmind.geoip', name: 'geoip-api', version:
                                                            '1.3.1'
    compile group: 'com.fasterxml.jackson.core', name: 'jackson-core',
version: '2.9.7'
}
jar {
  manifest {
```

```
    attributes 'Main-Class': mainClassName
  } from {
    configurations.compile.collect {
      it.isDirectory() ? it : zipTree(it)
    }
  }
  exclude "META-INF/*.SF"
  exclude "META-INF/*.DSA"
  exclude "META-INF/*.RSA"
}
```

Listing 3.1: build.gradle

Note that the first change is the switch from version 0.1.0 to version 0.2.0 .

The second change is to add the MaxMind's GeoIP version 1.3.1 to our project.

From the project root directory, run the following command to rebuild the app:

```
$ gradle jar
```

The output is something like the following:

```
...BUILD SUCCESSFUL in 8s
2 actionable tasks: 2 executed
```

To download a copy of the MaxMind GeoIP free database, execute this command:

```
$ wget
"http://geolite.maxmind.com/download/geoip/database/GeoLiteCity.dat.gz"
```

Run the following command to decompress the file:

```
$ gunzip GeoLiteCity.dat.gz
```

Move the GeoLiteCity.dat file in a route accessible to our program.

Now, add a file called GeoIPService.java in the src/main/java/monedero/extractors directory containing the content of *Listing 3.2*:

```
package monedero.extractors;
import com.maxmind.geoip.Location;
import com.maxmind.geoip.LookupService;
import java.io.IOException;
import java.util.logging.Level;
import java.util.logging.Logger;
public final class GeoIPService {
```

```
    private static final String MAXMINDDB =
"/path_to_your_GeoLiteCity.dat_file";
    public Location getLocation(String ipAddress) {
      try {
        final LookupService maxmind =
          new LookupService(MAXMINDDB, LookupService.GEOIP_MEMORY_CACHE);
        return maxmind.getLocation(ipAddress);
      } catch (IOException ex) {
        Logger.getLogger(GeoIPService.class.getName()).log(Level.SEVERE,
null, ex);
      }
      return null;
    }
}
```

Listing 3.2: GeoIPService.java

The GeoIPService has a public method getLocation that receives a string representing the IP address and looks for this IP address in the GeoIP location database. This method returns an object of class location with the geographic location of that specific IP address.

There are sometimes demanding customers who ask to have the most updated version of the database. In this case, downloading the database continuously is not an option. For this type of case, MaxMind exposes its services through an API. To read more about it, visit the following URL: https://dev.maxmind.com/.

To read more about Bit License regulations, visit the following link:

http://www.dfs.ny.gov/legal/regulations/bitlicense_reg_framework.html

Enriching the messages

Now, we will recap the steps of our processing engine for Monedero. The customer consults the ETH price in the client's browser and is sent to Kafka through some HTTP event collector.

The first step in our flow is the event correctness validation; remember from the previous chapter that the messages with defects are derived from bad data and that is why they are filtered. The second step now is to enrich our message with geographic location information.

Here are the architecture steps for the Monedero processing engine:

1. Read the individual events from a Kafka topic called **input-messages**
2. Validate the message, sending any defective event to a dedicated Kafka topic called **invalid-messages**
3. Enrich the message with the geographic location data
4. Write the enriched messages in a Kafka topic called **valid-messages**

All of these steps of the second version of the stream processing engine are detailed in *Figure 3.1*:

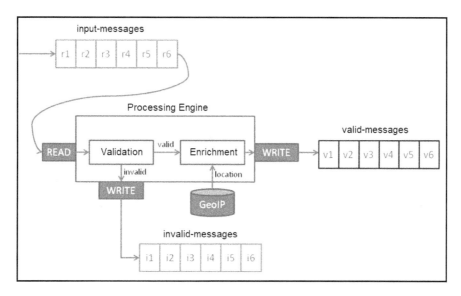

Figure 3.1: The processing engine reads the events from the input-messages topic, validates the messages, sends the errors to invalid-messages topic, enriches the messages with geographic location, and then writes them to the valid-messages topic.

Now, let's create a file called `Enricher.java` in the `src/main/java/monedero/` directory with the content of *Listing 3.3*:

```
package monedero;

import com.fasterxml.jackson.databind.JsonNode;
import com.fasterxml.jackson.databind.ObjectMapper;
import com.fasterxml.jackson.databind.node.ObjectNode;
import com.maxmind.geoip.Location;
import monedero.extractors.GeoIPService;
import org.apache.kafka.clients.producer.KafkaProducer;
import java.io.IOException;
```

```java
public final class Enricher implements Producer {
  private final KafkaProducer<String, String> producer;
  private final String validMessages;
  private final String invalidMessages;
  private static final ObjectMapper MAPPER = new ObjectMapper();
  public Enricher(String servers, String validMessages, String
    invalidMessages) {
    this.producer = new KafkaProducer<>
    (Producer.createConfig(servers));
    this.validMessages = validMessages;
    this.invalidMessages = invalidMessages;
  }
  @Override
  public void process(String message) {
    try {
      // this method below is filled below
    } catch (IOException e) {
      Producer.write(this.producer, this.invalidMessages, "{\"error\": \""
          + e.getClass().getSimpleName() + ": " + e.getMessage() + "\"}");
    }
  }
}
```

As expected, the `Enricher` class implements the producer interface; therefore the `Enricher` is a producer.

Let's fill the code of the `process()` method.

If the customer message does not have an IP address, the message is automatically sent to `invalid-messages` topic, as follows:

```java
final JsonNode root = MAPPER.readTree(message);
final JsonNode ipAddressNode =
  root.path("customer").path("ipAddress");
if (ipAddressNode.isMissingNode()) {
  Producer.write(this.producer, this.invalidMessages,
      "{\"error\": \"customer.ipAddress is missing\"}");
} else {
  final String ipAddress = ipAddressNode.textValue();
```

The `Enricher` class invokes the `getLocation` method of `GeoIPService`, as follows:

```java
final Location location = new GeoIPService().getLocation(ipAddress);
```

The country and the city of the location are added to the customer message, as in the example:

```
((ObjectNode) root).with("customer").put("country",
    location.countryName);
((ObjectNode) root).with("customer").put("city",
    location.city);
```

The enriched message is written to the `valid-messages` queue, as follows:

```
Producer.write(this.producer, this.validMessages,
    MAPPER.writeValueAsString(root));
}
```

Note that the location object brings more interesting data; for this example, just the city and the country are extracted. For example, the MaxMind database can give us much more precision than the one exploited in this example. In effect, the online API can accurately show the exact location of an IP.

Also note that here we have a very simple validation. In the next chapter, we will see how to validate the schema correctness. For the moment, think of other validations that are missing to have a system that meets the business requirements.

Extracting the currency price

At the moment, Monedero has a service that validates the messages that are well formed. The service also enriches the messages with the customer's geographic location.

Recall that the Monedero core business is the cryptocurrencies exchange. So now, the business asks us for a service that returns the requested currency price online at a specific time.

To achieve this, we will use the exchange rate of open exchange rates:

`https://openexchangerates.org/`

To obtain a free API key, you have to register in a free plan; the key is needed to access the free API.

Now, let's create a file called `OpenExchangeService.java` in the
`src/main/java/monedero/extractors` directory with the content of *Listing 3.4*:

```
package monedero.extractors;
import com.fasterxml.jackson.databind.JsonNode;
import com.fasterxml.jackson.databind.ObjectMapper;
import java.io.IOException;
import java.net.URL;
import java.util.logging.Level;
import java.util.logging.Logger;
public final class OpenExchangeService {
  private static final String API_KEY = "YOUR_API_KEY_VALUE_HERE";   //1
  private static final ObjectMapper MAPPER = new ObjectMapper();
  public double getPrice(String currency) {
    try {
      final URL url = new
URL("https://openexchangerates.org/api/latest.json?app_id=" + API_KEY);
//2
      final JsonNode root = MAPPER.readTree(url);
      final JsonNode node = root.path("rates").path(currency);   //3
      return Double.parseDouble(node.toString());                //4
    } catch (IOException ex) {
    Logger.getLogger(OpenExchangeService.class.getName()).log(Level.SEVERE,
null, ex);
    }
    return 0;
  }
}
```

Some lines of the `OpenExchangeService` class can be analyzed as follows:

- In line `//1`, the value of the `API_KEY` is assigned when you registered in the open
  exchange rates page; the free plan gives you up to 1,000 requests per month.
- In line `//2`, our class invokes the open exchange API URL, using your `API_KEY`.
  To check the prices at the moment, you can access the URL (counts as a request
  with your key): `https://openexchangerates.org/api/latest.json?app_id=`
  `YOUR_API_KEY`.
- In line `//3`, the currency string passed as argument is searched in the JSON tree
  that returns the web page.
- In line `//4`, the currency price (in US dollars) of the currency passed as an
  argument is returned as a double value.

There are several ways to parse JSON, and whole books are devoted to this topic. For this example, we used Jackson to parse JSON. To find more information, go to the following URL:

```
https://github.com/FasterXML
```

As with the MaxMind geographic localization service, open exchange rates also expose their services through an API. To read more about this, go to the following URL:

```
https://docs.openexchangerates.org/
```

This example uses the open exchange rates free plan; if a non-limited API is required, check their other plans in the URL:

```
https://openexchangerates.org/signup
```

Enriching with currency price

The customer consults the ETH price event, starts in the client's web browser, and is dispatched to Kafka through some HTTP event collector. The second step is to enrich the messages with the geographic location information from MaxMind database. The third step is to enrich the message with the currency price from open exchange rates service.

In summary, here are the architecture steps for the Monedero processing engine:

1. Read the individual events from a Kafka topic called **input-messages**
2. Validate the message, sending any defective event to a specific Kafka topic called **invalid-messages**
3. Enrich the message with the geographic location data from MaxMind database
4. Enrich the message with the currency price from open exchange rates service
5. Write the enriched messages in a Kafka topic called **valid-messages**

The final version of the stream processing engine is detailed in *Figure 3.2*:

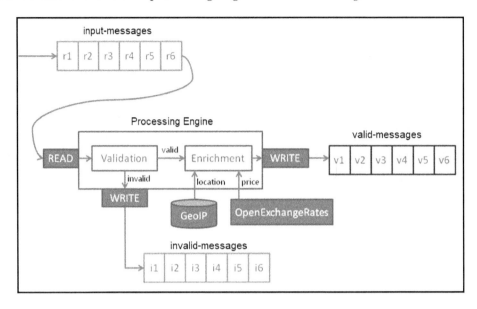

Figure 3.2: The processing engine reads the messages from the input-messages topic, validates the messages, routes the defective ones to invalid-messages queue, enriches the messages with geographic location and price, and finally, writes them to valid-messages queue.

To add the open exchange rates service to our engine, modify the Enricher.java file in the src/main/java/monedero/ directory with the changes highlighted in *Listing 3.5*:

```java
package monedero;
import com.fasterxml.jackson.databind.JsonNode;
import com.fasterxml.jackson.databind.ObjectMapper;
import com.fasterxml.jackson.databind.node.ObjectNode;
import com.maxmind.geoip.Location;
import monedero.extractors.GeoIPService;
import monedero.extractors.OpenExchangeService; //1
import org.apache.kafka.clients.producer.KafkaProducer;
import java.io.IOException;
public final class Enricher implements Producer {
  private final KafkaProducer<String, String> producer;
  private final String validMessages;
  private final String invalidMessages;
  private static final ObjectMapper MAPPER = new ObjectMapper();
  public Enricher(String servers, String validMessages, String
invalidMessages) {
    this.producer = new KafkaProducer<>(Producer.createConfig(servers));
    this.validMessages = validMessages;
    this.invalidMessages = invalidMessages;
  }
```

```
  @Override
  public void process(String message) {
    try {
      final JsonNode root = MAPPER.readTree(message);
      final JsonNode ipAddressNode =
root.path("customer").path("ipAddress");
      if (ipAddressNode.isMissingNode()) { //2
        Producer.write(this.producer, this.invalidMessages,
          "{\"error\": \"customer.ipAddress is missing\"}");
      } else {
        final String ipAddress = ipAddressNode.textValue();
        final Location location = new
GeoIPService().getLocation(ipAddress);
        ((ObjectNode) root).with("customer").put("country",
location.countryName);
        ((ObjectNode) root).with("customer").put("city", location.city);
        final OpenExchangeService oes = new OpenExchangeService(); //3
        ((ObjectNode) root).with("currency").put("rate",
oes.getPrice("BTC"));//4
        Producer.write(this.producer, this.validMessages,
MAPPER.writeValueAsString(root)); //5
      }
    } catch (IOException e) {
      Producer.write(this.producer, this.invalidMessages, "{\"error\": \""
        + e.getClass().getSimpleName() + ": " + e.getMessage() + "\"}");
    }
  }
}
```

As we know, the `Enricher` class is a Kafka producer, so now let's analyze the additions:

- In line `//1`, we import `OpenExchangeService` built previously
- In line `//2`, to avoid later null pointer exceptions, if the message does not have a valid IP Address on customer, the message automatically is sent to the `invalid-messages` queue
- In line `//3`, generates an instance of the `OpenExchangeService` class that is an extractor
- In line `//4`, the `getPrice()` method of the `OpenExchangeService` class is called, and this value is added to the message: the price of the currency is added to the currency node in the leaf price
- In line `//5`, the enriched message is written to the `valid-messages` queue

This is the final version of the enricher engine for Monedero; as we can see, the pipeline architecture uses the extractors as input for the enricher. Next, we will see how to run our entire project.

Note that the JSON response has a lot of more information, but for this example, only one currency price is used. There are several open data initiatives that are free and provide a lot of free repositories with online and historical data.

Running the engine

Now that the final version of the `Enricher` class is coded, we have to compile and execute it.

As we know, the `ProcessingEngine` class contains the main method to coordinate the reader and writer classes. Now, let's modify the `ProcessingEngine.java` file on the `src/main/java/monedero/` directory and replace `Validator` with `Enricher` as in the highlighted code in *Listing 3.6*:

```
package monedero;
public class ProcessingEngine {
  public static void main(String[] args){
    String servers = args[0];
    String groupId = args[1];
    String sourceTopic = args[2];
    String validTopic = args[3];
    String invalidTopic = args[4];
    Reader reader = new Reader(servers, groupId, sourceTopic);
    Enricher enricher = new Enricher(servers, validTopic, invalidTopic);
    reader.run(enricher);
  }
}
```

Listing 3.6: ProcessingEngine.java

The processing engine receives the following five arguments from the command line:

- `args[0] servers` indicates the host and port of the Kafka broker
- `args[1] groupId` indicates that the consumer is part of this Kafka consumer group
- `args[2] input topic` indicates the topic where the reader reads from
- `args[3] validTopic` indicates the topic where valid messages are sent
- `args[4] invalidTopic` indicates the topic where invalid messages are sent

To rebuild the project from the `monedero` directory, run the following command:

```
$ gradle jar
```

If everything is OK, the output should be similar to the following:

```
. . .
BUILD SUCCESSFUL in 8s
2 actionable tasks: 2 executed
```

To run the project, we need four different command-line windows. *Figure 3.3* shows the command-line windows arrangement:

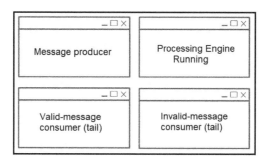

Figure 3.3: The four terminal windows to test the processing engine including: message producer, valid message consumer, invalid message consumer, and the processing engine itself

1. In the first command-line terminal, go to the Kafka installation directory and generate the two necessary topics, as follows:

```
$ bin/kafka-topics --create --zookeeper localhost:2181 --
replication-factor 1 --
partitions 1 --topic valid-messages
```

```
$ bin/kafka-topics --create --zookeeper localhost:2181 --
replication-factor 1 --
partitions 1 --topic invalid-messages
```

Then, start a console producer to the `input-topic` topic, as follows:

```
$ bin/kafka-console-producer --broker-list localhost:9092 --topic
input-topic
```

This window is where the input messages are produced (typed).

2. In the second command-line window, start a command-line consumer listening to the `valid-messages` topic, as follows:

```
$ bin/kafka-console-consumer --bootstrap-server localhost:9092 --
from-beginning -
-topic valid-messages
```

3. In the third command-line window, start a command-line consumer listening to `invalid-messages` topic, as follows:

```
$ bin/kafka-console-consumer --bootstrap-server localhost:9092 --
from-beginning -
-topic invalid-messages
```

4. In the fourth command-line terminal, start up the processing engine. From the project root directory (where the `gradle jar` command were executed) run this command:

```
$ java -jar ./build/libs/monedero-0.2.0.jar localhost:9092 foo
input-topic valid-
messages invalid-messages
```

From the first command-line terminal (the console producer), send the following three messages (remember to type enter between messages and execute each one in just one line):

```
{"event": "CUSTOMER_CONSULTS_ETHPRICE", "customer": {"id": "14862768",
"name": "Snowden, Edward", "ipAddress": "95.31.18.111"}, "currency":
{"name": "ethereum", "price": "USD"}, "timestamp": "2018-09-28T09:09:09Z"}
{"event": "CUSTOMER_CONSULTS_ETHPRICE", "customer": {"id": "13548310",
"name": "Assange, Julian", "ipAddress": "185.86.151.11"}, "currency":
{"name": "ethereum", "price": "USD"}, "timestamp": "2018-09-28T08:08:14Z"}
{"event": "CUSTOMER_CONSULTS_ETHPRICE", "customer": {"id": "15887564",
"name": "Mills, Lindsay", "ipAddress": "186.46.129.15"}, "currency":
{"name": "ethereum", "price": "USD"}, "timestamp": "2018-09-28T19:51:35Z"}
```

As these are valid messages, the messages typed in the producer console should appear in the valid-messages consumer console window, as in the example:

```
{"event": "CUSTOMER_CONSULTS_ETHPRICE", "customer": {"id": "14862768",
"name": "Snowden, Edward", "ipAddress": "95.31.18.111", "country":"Russian
Federation","city":"Moscow"}, "currency": {"name": "ethereum", "price":
"USD", "rate":0.0049}, "timestamp": "2018-09-28T09:09:09Z"}
{"event": "CUSTOMER_CONSULTS_ETHPRICE", "customer": {"id": "13548310",
"name": "Assange, Julian", "ipAddress": "185.86.151.11", "country":"United
Kingdom","city":"London"}, "currency": {"name": "ethereum", "price": "USD",
"rate":0.049}, "timestamp": "2018-09-28T08:08:14Z"}
{"event": "CUSTOMER_CONSULTS_ETHPRICE", "customer": {"id": "15887564",
"name": "Mills, Lindsay", "ipAddress": "186.46.129.15",
"country":"Ecuador","city":"Quito"}, "currency": {"name": "ethereum",
"price": "USD", "rate":0.049}, "timestamp": "2018-09-28T19:51:35Z"}
```

Extracting the weather data

Obtaining the geographic location from the IP address is a problem that has already been solved in this chapter.

In this last section, we will build another extractor that will be used in the following chapters. Now, suppose we want to know the current temperature of a given a geographic location at a specific time. To achieve this, we use the OpenWeatherService.

Visit the Open Weather page: `https://openweathermap.org/`.

To obtain a free API key register in a free plan; this key is needed to access the free API.

Now, create a file called `OpenWeatherService.java` in the `src/main/java/monedero/extractors` directory with the content of *Listing 3.7*:

```
package monedero.extractors;
import com.fasterxml.jackson.databind.JsonNode;
import com.fasterxml.jackson.databind.ObjectMapper;
import java.io.IOException;
import java.net.URL;
import java.util.logging.Level;
import java.util.logging.Logger;
public class OpenWeatherService {
  private static final String API_KEY = "YOUR API_KEY_VALUE"; //1
  private static final ObjectMapper MAPPER = new ObjectMapper();
  public double getTemperature(String lat, String lon) {
    try {
      final URL url = new URL(
          "http://api.openweathermap.org/data/2.5/weather?lat=" + lat
+ "&lon="+ lon +
          "&units=metric&appid=" + API_KEY); //2
      final JsonNode root = MAPPER.readTree(url);
      final JsonNode node = root.path("main").path("temp");/73
      return Double.parseDouble(node.toString());
    } catch (IOException ex) {
Logger.getLogger(OpenWeatherService.class.getName()).log(Level.SEVERE,
null, ex);
    }
    return 0;
  }
}
```

Listing 3.7: OpenWeatherService.java

The public method, `getTemperature()`, in the `OpenWeatherService` class receives two string values—the geographic latitude and longitude—and returns the current temperature for these locations. If the metric system is specified, the result will be in degrees celsius.

In a nutshell, the file includes the following:

- In line `//1`, to use the Open Weather API, a KEY is needed, registration is free, and gives 1,000 requests per month
- In line `//2`, to check the current weather at a particular location, open the following URL: `http://api.openweathermap.org/data/2.5/weather? lat=LAT lon=LONunits=metricappid=YOUR_API_KEY`
- In line `//3`, the JSON returned by this URL is parsed looking for the temperature

Open Weather also exposes their services through an API. To read how to use this API, go to the following:

`https://openweathermap.org/api`

Summary

In this chapter, we covered how to make data extraction, how message enrichment works, and how to extract the geographic location given an IP Address. Also, we demonstrated an example of how to extract the currency price given a currency and running a processing engine.

The `Chapter 4`, *Serialization*, talks about the schema registry. The extractors built in this chapter are used in the following chapters.

4
Serialization

In modern (internet) computing, we often forget that entities must be transmitted from one computer to another. In order to be able to transmit the entities, they must first be serialized.

Serialization is the process of transforming an object into a stream of bytes commonly used to transmit it from one computer to another.

Deserialization, as the name implies, is the opposite of serialization, that is, to convert a stream of bytes into an object (for didactic purposes, we can say that the object is inflated or rehydrated), normally from the side that receives the message. Kafka provides **Serializer/Deserializer** (**SerDe**) for the primitive data types (byte, integer, long, double, String, and so on).

In this chapter, a new company is introduced: Kioto (standing for Kafka Internet of Things). This chapter covers the following topics:

- How to build a Java `PlainProducer`, a consumer, and a processor
- How to run a Java `PlainProducer` and a processor
- How to build a custom serializer and a custom deserializer
- How to build a Java `CustomProducer`, a consumer, and a processor
- How to run a Java `CustomProducer` and a processor

Kioto, a Kafka IoT company

Kioto is a fictional company dedicated to energy production and distribution. To operate, Kioto has several **Internet of Things** (**IoT**) devices.

Kioto also wants to build an enterprise service bus with Apache Kafka. Its goal is to manage all of the messages received by all of the machines' IoT sensors every minute. Kioto has hundreds of machines in several locations, sending thousands of different messages per minute to the enterprise service bus.

As mentioned, Kioto has a lot of IoT on its machines that continuously send status messages to a control center. These machines generate electricity, so it is very important for Kioto to know exactly the machines' uptime and their state (running, shutting down, shutdown, starting, and so on).

Kioto also needs to know the weather forecast, because some machines should not operate over certain temperatures. Some machines display different behavior based on the environmental temperature. It is different starting a machine in cold rather than in warm conditions, so the start up time is important when calculating the uptime. To warrant the continuous electricity supply, the information has to be precise. It is always better to face an electrical power failure having to start the machines from a warm temperature rather than from cold temperature.

Listing 4.1 shows the health check event in JSON format.

The following is the content of *Listing 4.1*, `healthcheck.json`:

```
{
    "event":"HEALTH_CHECK",
    "factory":"Duckburg",
    "serialNumber":"R2D2-C3P0",
    "type":"GEOTHERMAL",
    "status":"RUNNING",
    "lastStartedAt":"2017-09-04T17:27:28.747+0000",
    "temperature":31.5,
    "ipAddress":"192.166.197.213"}
}
```

Listing 4.1: healthcheck.json

The proposed representation of this message in JSON has the following properties:

- event: The string with the message's type (in this case, HEALTH_CHECK)
- factory: The name of the city where the plant is physically located
- serialNumber: The machine's serial number
- type: Represents the machine's type, which could be GEOTHERMAL, HYDROELECTRIC, NUCLEAR, WIND, or SOLAR
- status: The point on the life cycle: RUNNING, SHUTTING-DOWN, SHUT-DOWN, STARTING
- lastStartedAt: The last start time
- temperature: A float representing the machine's temperature in degrees celsius
- ipAddress: The machine's IP address

As we can see, JSON is a human-readable message format.

Project setup

The first step is to create the Kioto project. Create a directory called `kioto`. Go to that directory and execute the following command:

```
$ gradle init --type java-library
```

The output is something like the following:

```
Starting a Gradle Daemon (subsequent builds will be faster)
BUILD SUCCESSFUL in 3s
2 actionable tasks: 2 execute BUILD SUCCESSFUL
```

Gradle creates a default project in the directory, including two Java files called `Library.java` and `LibraryTest.java`; delete both files.

Your directory should be similar to the following:

- - build.gradle
- - gradle
- -- wrapper
- --- gradle-wrapper.jar
- --- gradle-vreapper.properties
- - gradlew
- - gradle.bat
- - settings.gradle
- - src
- -- main
- --- java
- ----- Library.java
- -- test
- --- java
- ----- LibraryTest.java

Modify the `build.gradle` file and replace it with *Listing 4.2*.

The following is the content of *Listing 4.2*, the Kioto Gradle build file:

```
apply plugin: 'java'
apply plugin: 'application'
sourceCompatibility = '1.8'
mainClassName = 'kioto.ProcessingEngine'
repositories {
    mavenCentral()
    maven { url 'https://packages.confluent.io/maven/' }
}
version = '0.1.0'
dependencies {
    compile 'com.github.javafaker:javafaker:0.15'
    compile 'com.fasterxml.jackson.core:jackson-core:2.9.7'
    compile 'io.confluent:kafka-avro-serializer:5.0.0'
    compile 'org.apache.kafka:kafka_2.12:2.0.0'
}
jar {
    manifest {
        attributes 'Main-Class': mainClassName
    } from {
        configurations.compile.collect {
            it.isDirectory() ? it : zipTree(it)
        }
    }
    exclude "META-INF/*.SF"
    exclude "META-INF/*.DSA"
    exclude "META-INF/*.RSA"
}
```

Some library dependencies added to the application are as follows:

- `kafka_2.12`, the necessary dependencies for Apache Kafka
- `javafaker`, the necessary dependencies for JavaFaker
- `jackson-core`, for JSON parsing and manipulation
- `kafka-avro-serializer`, to serialize in Kafka with Apache Avro

Note that to use the `kafka-avro-serializer` function, we added the Confluent repository in the repositories section.

To compile the project and download the required dependencies, type the following command:

```
$ gradle compileJava
```

The output should be similar to the following:

```
BUILD SUCCESSFUL in 3s
1 actionable task: 1 executed
```

The project can also be created with Maven, SBT, or even from the IDE. But for simplicity, it was created with Gradle. For more information about these projects, visit the following:

- Gradle's main page: http://www.gradle.org
- Maven's main page: http://maven.apache.org
- SBT's main page: http://www.scala-sbt.org/
- Jackson's main page: https://github.com/FasterXML
- JavaFaker's main page: https://github.com/DiUS/java-faker

The constants

The first step is to code our `Constants` class. This class is a static class with all of the `Constants` needed in our project.

Open the project with your favorite IDE and, under the `src/main/java/kioto` directory, create a file called `Constants.java` with the content of *Listing 4.3*.

The following is the content of *Listing 4.3*, `Constants.java`:

```java
package kioto;
import com.fasterxml.jackson.databind.ObjectMapper;
import com.fasterxml.jackson.databind.SerializationFeature;
import com.fasterxml.jackson.databind.util.StdDateFormat;
public final class Constants {
  private static final ObjectMapper jsonMapper;
  static {
    ObjectMapper mapper = new ObjectMapper();
    mapper.disable(SerializationFeature.WRITE_DATES_AS_TIMESTAMPS);
    mapper.setDateFormat(new StdDateFormat());
    jsonMapper = mapper;
  }
  public static String getHealthChecksTopic() {
    return "healthchecks";
  }
  public static String getHealthChecksAvroTopic() {
    return "healthchecks-avro";
  }
  public static String getUptimesTopic() {
    return "uptimes";
```

```
  }
  public enum machineType {GEOTHERMAL, HYDROELECTRIC, NUCLEAR, WIND, SOLAR}
  public enum machineStatus {STARTING, RUNNING, SHUTTING_DOWN, SHUT_DOWN}
  public static ObjectMapper getJsonMapper() {
    return jsonMapper;
  }
}
```

In our `Constants` class, there are some methods that we will need later. These are as follows:

- `getHealthChecksTopic`: It returns the name of the health checks input topic
- `getHealthChecksAvroTopic`: It returns the name of the topic with the health checks in Avro
- `getUptimesTopic`: It returns the name of the `uptimes` topic
- `machineType`: This is an `enum` with the types of the Kioto energy producing machines types
- `machineType`: This is an `enum` with the types of the Kioto machines' possible statuses
- `getJsonMapper`: It returns the object mapper for JSON serialization and we set the serialization format for dates

This is a `Constants` class; in languages such as Kotlin, the constants don't require an independent class, but we are using Java. Some purists of object-oriented programming argue that to code constant classes is an object-oriented anti-pattern. However, for simplicity here, we need some constants in our system.

HealthCheck message

The second step is to code the `HealthCheck` class. This class is a **Plain Old Java Object** (**POJO**). The `model` class is the template for the value object.

Open the project with your favorite IDE and, in the `src/main/java/kioto` directory, create a file called `HealthCheck.java` with the content of *Listing 4.4*.

The following is the content of *Listing 4.4*, `HealthCheck.java`:

```
package kioto;
import java.util.Date;
public final class HealthCheck {
  private String event;
  private String factory;
```

```
    private String serialNumber;
    private String type;
    private String status;
    private Date lastStartedAt;
    private float temperature;
    private String ipAddress;
}
```

Listing 4.4: HealthCheck.java

With your IDE, generate the following:

- A no-parameter constructor
- A constructor with all of the attributes passed as parameters
- The getters and the setters for each attribute

This is a data class, a POJO in Java. In languages such as Kotlin, the model classes require so much less boilerplate code, but now we are in Java. Some purists of object-oriented programming argue that value objects is an object-oriented anti-pattern. However, the serialization libraries to produce messages need these classes.

To generate fake data with JavaFaker, our code should be as shown in *Listing 4.5.*

The following is the content of *Listing 4.5*, a health check mock generator with JavaFaker:

```
HealthCheck fakeHealthCheck =
    new HealthCheck(
        "HEALTH_CHECK",
        faker.address().city(),                           //1
        faker.bothify("??##-??##", true),       //2
            Constants.machineType.values()
                [faker.number().numberBetween(0,4)].toString(), //3
        Constants.machineStatus.values()
                [faker.number().numberBetween(0,3)].toString(), //4
        faker.date().past(100, TimeUnit.DAYS),            //5
        faker.number().numberBetween(100L, 0L),           //6
        faker.internet().ipV4Address());                  //7
```

The following is an analysis of how to generate fake health check data:

- In line //1, address().city() generates a fictitious city name
- In line //2, in the expression ? for alpha # for numeric, true if alpha is uppercase
- In line //3, we use the machine type enum in Constants , and a fake number between 0 and 4

- In line `//4`, we use the machine status `enum` in `Constants` and a fake number between `0` and `3`, inclusively
- In line `//5`, we are saying that we want a fake date between the past `100` days from today
- In line `//6`, we build a fake IP address

Here, we depend on the attributes order of the constructor. Other languages, such as Kotlin, allow specifying each assigned attribute name.

Now, to transform our Java POJO into a JSON string, we use the method in the `Constants` class—something like the following:

```
String fakeHealthCheckJson fakeHealthCheckJson =
Constants.getJsonMapper().writeValueAsString(fakeHealthCheck);
```

Don't forget that this method throws a JSON processing exception.

Java PlainProducer

As we already know, to build a Kafka Message producer that we use the Java client library, in particular the producer API (in the following chapters, we will see how to use Kafka Streams and KSQL).

The first thing we need is a data source; to make it simple we need to produce our mock data. Each message will be a health message with all of its attributes. The first step is to build a producer to send these messages in JSON format to a topic, as in the example:

```
{"event":"HEALTH_CHECK","factory":"Port Roelborough","serialNumber":"QT89-
TZ50","type":"GEOTHERMAL","status":"SHUTTING_DOWN","lastStartedAt":"2018-09
-13T00:36:39.079+0000","temperature":28.0,"ipAddress":"235.180.238.3"}

{"event":"HEALTH_CHECK","factory":"Duckburg","serialNumber":"NB49-
XL51","type":"NUCLEAR","status":"RUNNING","lastStartedAt":"2018-08-18T05:42
:29.648+0000","temperature":49.0,"ipAddress":"42.181.105.188"}
...
```

Let's start by creating a Kafka producer that we will use to send the input messages.

As we already know, there are two requisites that all of the Kafka producers should have: they must be `KafkaProducer` and have specific properties set, as shown in *Listing 4.6*.

The following is the content of *Listing 4.6*, the constructor method for `PlainProducer`:

```
import org.apache.kafka.clients.producer.KafkaProducer;
import org.apache.kafka.clients.producer.Producer;
import org.apache.kafka.common.serialization.StringSerializer;
public final class PlainProducer {
  private final Producer<String, String> producer;
  public PlainProducer(String brokers) {
    Properties props = new Properties();
    props.put("bootstrap.servers", brokers);              //1
    props.put("key.serializer", StringSerializer.class);   //2
    props.put("value.serializer", StringSerializer.class); //3
    producer = new KafkaProducer<>(props);                 //4
  }
  . . .
}
```

An analysis of the `PlainProducer` constructor includes the following:

- In line `//1`, the list of the brokers where our producer will be running
- In line `//2`, the serializer type for the messages' keys (we will see serializers later)
- In line `//3`, the serializer type for the messages' values (in this case, the values are strings)
- In line `//4`, with these properties we build a `KafkaProducer` with string keys and string values, for example, `<String, String>`
- Note that properties behave like a HashMap; in languages such as Kotlin, the properties assignment could be made using the = operator, rather than by calling a method

We are using a string serializer for both keys and values: in this first approach, we will serialize the values to JSON manually using Jackson. We will see later how to write a custom serializer.

Now, in the `src/main/java/kioto/plain` directory, create a file called `PlainProducer.java` with the content of *Listing 4.7*.

The following is the content of *Listing 4.7*, `PlainProducer.java`:

```
package kioto.plain;
import ...
public final class PlainProducer {
  /* here the Constructor code in Listing 4.6 */
  public void produce(int ratePerSecond) {
    long waitTimeBetweenIterationsMs = 1000L / (long)ratePerSecond; //1
    Faker faker = new Faker();
```

```
      while(true) { //2
        HealthCheck fakeHealthCheck /* here the code in Listing 4.5 */;
        String fakeHealthCheckJson = null;
        try {
          fakeHealthCheckJson =
  Constants.getJsonMapper().writeValueAsString(fakeHealthCheck); //3
        } catch (JsonProcessingException e) {
          // deal with the exception
        }
        Future futureResult = producer.send(new ProducerRecord<>
          (Constants.getHealthChecksTopic(), fakeHealthCheckJson)); //4
        try {
          Thread.sleep(waitTimeBetweenIterationsMs); //5
          futureResult.get(); //6
        } catch (InterruptedException | ExecutionException e) {
          // deal with the exception
        }
      }
    }
    public static void main(String[] args) {
      new PlainProducer("localhost:9092").produce(2); //7
    }
  }
```

An analysis of the `PlainProducer` class includes the following:

- In line `//1`, `ratePerSecond` is the number of messages to send in a one second period
- In line `//2`, to simulate repetition, we use an infinite loop (try to avoid this in production)
- In line `//3`, the code to serialize as JSON a Java POJO
- In line `//4`, we use a Java Future to send the message to `HealthChecksTopic`
- In line `//5`, we wait this time to send messages again
- In line `//6`, we read the result of the future created previously
- In line `//7`, everything runs on the broker in localhost in port 9092, sending two messages at intervals of one second

It is important to note that here we are sending records without a key; we only specified the value (a JSON string), so the key is `null`. We are also calling the `get()` method on the result in order to wait for the write acknowledgment: without that, messages could be sent to Kafka but are lost without our program noticing the failure.

Running the PlainProducer

To build the project, run this command from the `kioto` directory:

```
$ gradle jar
```

If everything is okay, the output is something like the following:

```
BUILD SUCCESSFUL in 3s
1 actionable task: 1 executed
```

1. From a command-line terminal, move to the `confluent` directory and start it by typing the following:

```
$ ./bin/confluent start
```

2. The broker is running on port `9092`. To create the `healthchecks` topic, execute the following:

```
$ ./bin/kafka-topics --zookeeper localhost:2181 --create --topic
healthchecks --replication-factor 1 --partitions 4
```

3. Run a console consumer for the `healthchecks` topic by typing the following:

```
$ ./bin/kafka-console-consumer --bootstrap-server localhost:9092
--topic healthchecks
```

4. From our IDE, run the main method of the `PlainProducer`

5. The output on the console consumer should be similar to the following:

```
{"event":"HEALTH_CHECK","factory":"Lake
Anyaport","serialNumber":"EW05-
HV36","type":"WIND","status":"STARTING","lastStartedAt":"2018-09-17
T11:05:26.094+0000","temperature":62.0,"ipAddress":"15.185.195.90"}

{"event":"HEALTH_CHECK","factory":"Candelariohaven","serialNumber":
"BO58-
SB28","type":"SOLAR","status":"STARTING","lastStartedAt":"2018-08-1
6T04:00:00.179+0000","temperature":75.0,"ipAddress":"151.157.164.16
2"}

{"event":"HEALTH_CHECK","factory":"Ramonaview","serialNumber":"DV03
-
ZT93","type":"SOLAR","status":"RUNNING","lastStartedAt":"2018-07-12
T10:16:39.091+0000","temperature":70.0,"ipAddress":"173.141.90.85"}
...
```

Remember that, when producing data, there are several write guarantees that we could achieve.

For example, in case of a network failure or a broker failure, is our system ready to lose data?

There is a trade-off among three factors: the availability to produce messages, the latency in the production, and the guarantee of the safe write.

In this example, we just have one broker, and we use the default value for `acks` of 1. When we call the `get()` method in the future, we are waiting for the broker acknowledgment, that is, we have a guarantee that the message is persisted before sending another message. In this configuration, we don't lose messages, but our latency is higher than in a fire and forget schema.

Java plain consumer

As we already know, to build a Kafka message consumer, we use the Java client library—in particular, the consumer API (in the following chapters, we will see how to use Kafka Streams and KSQL).

Let's create a Kafka consumer that we will use to receive the input messages.

As we already know, there are two requisites that all of the Kafka consumers should have: to be a `KafkaConsumer` and to set the specific properties, such as those shown in *Listing 4.8*.

The following is the content of *Listing 4.8*, the constructor method for plain consumer:

```
import org.apache.kafka.clients.consumer.KafkaConsumer;
import org.apache.kafka.clients.consumer.Consumer;
import org.apache.kafka.common.serialization.StringSerializer;
public final class PlainConsumer {
  private Consumer<String, String> consumer;
  public PlainConsumer(String brokers) {
    Properties props = new Properties();
    props.put("group.id", "healthcheck-processor");        //1
    props.put("bootstrap.servers", brokers);               //2
    props.put("key.deserializer", StringDeserializer.class);   //3
    props.put("value.deserializer", StringDeserializer.class); //4
    consumer = new KafkaConsumer<>(props);                  //5
  }
  ...
}
```

An analysis of the plain consumer constructor includes the following:

- In line `//1`, the group ID of our consumer, in this case, `healthcheck-processor`
- In line `//2`, the list of `brokers` where our consumer will be running
- In line `//3`, the deserializer type for the messages' keys (we will see deserializers later)
- In line `//4`, the deserializer type for the messages' values, in this case, values are strings
- In line `//5`, with these properties, we build a `KafkaConsumer` with string keys and string values, for example, `<String, String>`

For the customers, we need to provide a group ID to specify the consumer group that our consumer will join.

In the case that multiple consumers are started in parallel, through different threads or through different processes, each consumer will be assigned with a subset of the topic partitions. In our example, we created our topic with four partitions, which means that, to consume the data in parallel, we could create up to four consumers.

For a consumer, we provide deserializers rather than serializers. Although we don't use the key deserializer (because if you remember, it is `null`), the key deserializer is a mandatory parameter for the consumer specification. On the other hand, we need the deserializer for the value, because we are reading our data in a JSON string, whereas here we deserialize the object manually with Jackson.

Java PlainProcessor

Now, in the `src/main/java/kioto/plain` directory, create a file called `PlainProcessor.java` with the content of *Listing 4.9*.

The following is the content of *Listing 4.9*, `PlainProcessor.java` (part 1):

```
package kioto.plain;
import ...
public final class PlainProcessor {
  private Consumer<String, String> consumer;
  private Producer<String, String> producer;
  public PlainProcessor(String brokers) {
    Properties consumerProps = new Properties();
    consumerProps.put("bootstrap.servers", brokers);
    consumerProps.put("group.id", "healthcheck-processor");
```

```
consumerProps.put("key.deserializer", StringDeserializer.class);
consumerProps.put("value.deserializer", StringDeserializer.class);
consumer = new KafkaConsumer<>(consumerProps);
Properties producerProps = new Properties();
producerProps.put("bootstrap.servers", brokers);
producerProps.put("key.serializer", StringSerializer.class);
producerProps.put("value.serializer", StringSerializer.class);
producer = new KafkaProducer<>(producerProps);
}
```

An analysis of the first part of the `PlainProcessor` class includes the following:

- In the first part, we declare a consumer, as in *Listing 4.8*
- In the second part, we declare a producer, as in *Listing 4.6*

Before continuing to write code, let's remember the project requirements for the Kioto stream processing engine.

Putting it all together, the specification is to create a stream engine that does the following:

- Generates messages to a Kafka topic called **healthchecks**
- Reads messages from the Kafka topic called **healthchecks**
- Calculates the uptime based on the start up time
- Writes the messages in a Kafka topic called **uptimes**

This entire process is detailed in *Figure 4.1*, that is, the Kioto stream processing application:

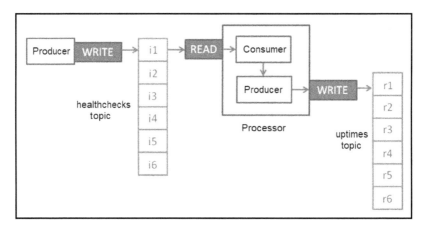

Figure 4.1: The messages are generated into HealthChecksTopic, then read, and finally the calculated uptimes are written it in the uptimes topic.

Now that we're in the `src/main/java/kioto/plain` directory, let's complete the `PlainProcessor.java` file with the content of *Listing 4.10*.

The following is the content of *Listing 4.10*, `PlainProcessor.java` (part 2):

```
public final void process() {
    consumer.subscribe(Collections.singletonList(
            Constants.getHealthChecksTopic()));         //1
    while(true) {
      ConsumerRecords records = consumer.poll(Duration.ofSeconds(1L)); //2
      for(Object record : records) {                  //3
        ConsumerRecord it = (ConsumerRecord) record;
        String healthCheckJson = (String) it.value();
        HealthCheck healthCheck = null;
        try {
          healthCheck = Constants.getJsonMapper()
           .readValue(healthCheckJson, HealthCheck.class);    // 4
        } catch (IOException e) {
          // deal with the exception
        }
        LocalDate startDateLocal
=healthCheck.getLastStartedAt().toInstant()
.atZone(ZoneId.systemDefault()).toLocalDate();       //5
        int uptime =
            Period.between(startDateLocal, LocalDate.now()).getDays();
//6
        Future future =
            producer.send(new ProducerRecord<>(
                        Constants.getUptimesTopic(),
                        healthCheck.getSerialNumber(),
                        String.valueOf(uptime)));
//7
        try {
          future.get();
        } catch (InterruptedException | ExecutionException e) {
          // deal with the exception
        }
      }
    }
  }
  public static void main( String[] args) {
    (new PlainProcessor("localhost:9092")).process();
  }
}
```

Listing 4.10: PlainProcessor.java (part 2)

An analysis of the `PlainProcessor` includes the following:

- In line `//1`, the consumer is created and subscribed to the source topic. This is a dynamic assignment of the partitions to our customer and join to the customer group.
- In line `//2`, an infinite loop to consume the records, the pool duration is passed as a parameter to the method pool. The customer waits no longer than one second before return.
- In line `//3`, we iterate over the records.
- In line `//4`, the JSON string is deserialized to extract the health check object.
- In line `//5`, the start time is transformed formatted at the current time zone.
- In line `//6`, the uptime is calculated.
- In line `//7`, the uptime is written to the `uptimes` topic, using the serial number as the key and the uptime as value. Both values are written as normal strings.

The moment at which the broker returns records to the client also depends on the `fetch.min.bytes` value; its default is 1, and is the minimum data amount to wait before the broker is available to the client. Our broker returns as soon as 1 byte of data is available, while waiting a maximum of one second.

The other configuration property is `fetch.max.bytes`, which defines the amount of data returned at once. With our configuration, the broker will return all of the available records (without exceeding the maximum of 50 MB).

If there are no records available, the broker returns an empty list.

Note that we could reuse the producer that generates the mock data, but it is clearer to use another producer to write `uptimes`.

Running the PlainProcessor

To build the project, run the following command from the `kioto` directory:

```
$ gradle jar
```

If everything is correct, the output is something like the following:

```
BUILD SUCCESSFUL in 3s
1 actionable task: 1 executed
```

1. Our broker is running on port `9092`, so to create the `uptimes` topic, execute the following command:

```
$ ./bin/kafka-topics --zookeeper localhost:2181 --create --topic
uptimes --replication-factor 1 --partitions 4
```

2. Run a console consumer for the `uptimes` topic, as follows:

```
$ ./bin/kafka-console-consumer --bootstrap-server localhost:9092
--topic uptimes --property print.key=true
```

3. From our IDE, run the main method of `PlainProcessor`
4. From our IDE, run the main method of `PlainProducer`
5. The output on the console consumer for the `uptimes` topic should be similar to the following:

```
EW05-HV36    33
BO58-SB28    20
DV03-ZT93    46
. . .
```

We have said that, when producing data, there are two factors to think about; one is the delivery guarantee, and the other is the partitioning.

When consuming data, we have to think about the following four factors:

- The number of consumers to run in parallel (in parallel threads and/or parallel processes)
- The amount of data to consume at once (think in terms of memory)
- The time to wait to receive messages (throughput and latency)
- When to mark a message as processed (committing offset)

If `enable.auto.commit` is set to `true` (the default is `true`), the consumer automatically will commit the offsets in the next call to the poll method.

Note that the whole batch of records is committed; if something fails and the application crashes after processing only some messages, but not all of the batch, the events are not committed and they will be reprocessed by other consumer; this way to process data is called at least once processing.

Custom serializer

So far, we have seen how to produce and consume JSON messages using plain Java and Jackson. We will see here how to create our custom serializers and deserializers.

We have seen how to use `StringSerializer` in the producer and `StringDeserializer` in the consumer. Now, we will see how to build our own SerDe to abstract the serialization/deserialization processes away from the core code of the application.

To build a custom serializer, we need to create a class that implements the `org.apache.kafka.common.serialization.Serializer` interface. This is a generic type, so we can indicate the custom type to be converted into an array of bytes (serialization).

In the `src/main/java/kioto/serde` directory, create a file called `HealthCheckSerializer.java` with the content of *Listing 4.11*.

The following is the content of *Listing 4.11*, `HealthCheckSerializer.java`:

```
package kioto.serde;
import com.fasterxml.jackson.core.JsonProcessingException;
import kioto.Constants;
import java.util.Map;
import org.apache.kafka.common.serialization.Serializer;
public final class HealthCheckSerializer implements Serializer {
  @Override
  public byte[] serialize(String topic, Object data) {
    if (data == null) {
      return null;
    }
    try {
      return Constants.getJsonMapper().writeValueAsBytes(data);
    } catch (JsonProcessingException e) {
      return null;
    }
  }

  @Override
  public void close() {}
  @Override
  public void configure(Map configs, boolean isKey) {}
}
```

Listing 4.11: HealthCheckSerializer.java

Note that the serializer class is located in a special module called **kafka-clients** in the `org.apache.kafka` route. The objective here is to use the serializer class instead of Jackson (manually).

Also note that the important method to implement is the `serialize` method. The `close` and `configure` methods can be left with an empty body.

We import the `JsonProcessingException` of Jackson just because the `writeValueAsBytes` method throws this exception, but we don't use Jackson for serialization.

Java CustomProducer

Now, to incorporate the serializer in our producer, there are two requisites that all Kafka producers should fulfill: to be a `KafkaProducer`, and to set the specific properties, such as *Listing 4.12*.

The following is the content of *Listing 4.12*, the constructor method for `CustomProducer`:

```
import kioto.serde.HealthCheckSerializer;
import org.apache.kafka.clients.producer.KafkaProducer;
import org.apache.kafka.clients.producer.Producer;
import org.apache.kafka.common.serialization.StringSerializer;
public final class CustomProducer {
  private final Producer<String, HealthCheck> producer;
  public CustomProducer(String brokers) {
    Properties props = new Properties();
    props.put("bootstrap.servers", brokers);              //1
    props.put("key.serializer", StringSerializer.class);   //2
    props.put("value.serializer", HealthCheckSerializer.class); //3
    producer = new KafkaProducer<>(props);                 //4
  }
```

An analysis of the `CustomProducer` constructor includes the following:

- In line `//1`, this is the list of the brokers where our producer will be running.
- In line `//2`, the serializer type for the messages' keys in this case keys remains as strings. In line `//3`, this is the serializer type for the messages' values, in this case, the values are `HealthCheck`.
- In line `//4`, with these properties we build a `KafkaProducer` with string keys and `HealthCheck` values, for example, `<String, HealthCheck>`.

Now, in the `src/main/java/kioto/custom` directory, create a file called `CustomProducer.java` with the content of *Listing 4.13*.

The following is the content of *Listing 4.13*, `CustomProducer.java`:

```
package kioto.plain;
import ...
public final class CustomProducer {
  /* here the Constructor code in Listing 4.12 */
  public void produce(int ratePerSecond) {
    long waitTimeBetweenIterationsMs = 1000L / (long)ratePerSecond; //1
    Faker faker = new Faker();
    while(true) { //2
      HealthCheck fakeHealthCheck /* here the code in Listing 4.5 */;
      Future futureResult = producer.send( new ProducerRecord<>(
        Constants.getHealthChecksTopic(), fakeHealthCheck));        //3
      try {
        Thread.sleep(waitTimeBetweenIterationsMs); //4
        futureResult.get();        //5
      } catch (InterruptedException | ExecutionException e) {
        // deal with the exception
      }
    }
  }
  public static void main(String[] args) {
    new CustomProducer("localhost:9092").produce(2); //6
  }
}
```

Listing 4.13: CustomProducer.java

An analysis of the `CustomProducer` class includes the following:

- In line `//1`, `ratePerSecond` is the number of messages to send in a one-second period
- In line `//2`, to simulate repetition, we use a infinite loop (try to avoid this in production)
- In line `//3`, we use a Java future to send the message to `HealthChecksTopic`
- In line `//4`, we wait this time to send messages again
- In line `//5`, we read the result of the future created previously
- In line `//6`, everything runs on the broker in localhost in port `9092`, sending two messages in an interval of one second

Running the CustomProducer

To build the project, run the following command from the `kioto` directory:

```
$ gradle jar
```

If everything is okay, the output is something like the following:

```
BUILD SUCCESSFUL in 3s
1 actionable task: 1 executed
```

1. Run a console consumer for `HealthChecksTopic` as follows:

```
$ ./bin/kafka-console-consumer --bootstrap-server localhost:9092
--topic healthchecks
```

2. From our IDE, run the main method of the `CustomProducer`
3. The output on the console consumer should be similar to the following:

```
{"event":"HEALTH_CHECK","factory":"Lake
Anyaport","serialNumber":"EW05-
HV36","type":"WIND","status":"STARTING","lastStartedAt":"2018-09-17
T11:05:26.094+0000","temperature":62.0,"ipAddress":"15.185.195.90"}

{"event":"HEALTH_CHECK","factory":"Candelariohaven","serialNumber":
"BO58-
SB28","type":"SOLAR","status":"STARTING","lastStartedAt":"2018-08-1
6T04:00:00.179+0000","temperature":75.0,"ipAddress":"151.157.164.16
2"}

{"event":"HEALTH_CHECK","factory":"Ramonaview","serialNumber":"DV03
-
ZT93","type":"SOLAR","status":"RUNNING","lastStartedAt":"2018-07-12
T10:16:39.091+0000","temperature":70.0,"ipAddress":"173.141.90.85"}

. . .
```

Custom deserializer

In a similar way, to build a custom deserializer, we need to create a class that implements the `org.apache.kafka.common.serialization.Deserializer` interface. We must indicate how to convert an array of bytes into a custom type (deserialization).

In the `src/main/java/kioto/serde` directory, create a file called `HealthCheckDeserializer.java` with the content of *Listing 4.14.*

The following is the content of *Listing 4.14,* `HealthCheckDeserializer.java`:

```
package kioto.serde;
import kioto.Constants;
import kioto.HealthCheck;
import java.io.IOException;
import java.util.Map;
import org.apache.kafka.common.serialization.Deserializer;
public final class HealthCheckDeserializer implements Deserializer {
  @Override
  public HealthCheck deserialize(String topic, byte[] data) {
    if (data == null) {
      return null;
    }
    try {
      return Constants.getJsonMapper().readValue(data, HealthCheck.class);
    } catch (IOException e) {
      return null;
    }
  }
  @Override
  public void close() {}
  @Override
  public void configure(Map configs, boolean isKey) {}
}
```

Listing 4.14: HealthCheckDeserializer.java

Note that the deserializer class is located in a module called kafka-clients in the `org.apache.kafka` route. The objective here is to use the deserializer class instead of Jackson (manually).

Also note that the important method to implement is the `deserialize` method. The `close` and `configure` methods can be left with an empty body.

We import the `HealthCheck` class because the `readValue` method requires a POJO (a class with public constructor and public getters and setters). Note also that all of the POJO attributes should be serializables.

Java custom consumer

Let's create a Kafka consumer that we will use to receive the custom input messages.

Now, in order to incorporate the deserializer in our consumer, there are two requisites that all of the Kafka consumers should have: to be a `KafkaConsumer`, and to set the specific properties, such as those in *Listing 4.15*.

The following is the content of *Listing 4.15*, the constructor method for `CustomConsumer`:

```
import kioto.HealthCheck;
import kioto.serde.HealthCheckDeserializer;
import org.apache.kafka.clients.consumer.KafkaConsumer;
import org.apache.kafka.clients.consumer.Consumer;
import org.apache.kafka.common.serialization.StringSerializer;
public final class CustomConsumer {
  private Consumer<String, HealthCheck> consumer;
  public CustomConsumer(String brokers) {
    Properties props = new Properties();
    props.put("group.id", "healthcheck-processor");//1
    props.put("bootstrap.servers", brokers);//2
    props.put("key.deserializer", StringDeserializer.class);//3
    props.put("value.deserializer", HealthCheckDeserializer.class); //4
    consumer = new KafkaConsumer<>(props);//5
  }
  ...
}
```

An analysis of the `CustomConsumer` constructor includes the following:

- In line `//1`, the group ID of our consumer, in this case, `healthcheck-processor`
- In line `//2`, the list of the brokers where our consumer will be running
- In line `//3`, the deserializer type for the messages' keys; in this case, the keys remains as strings
- In line `//4`, the deserializer type for the messages' values; in this case, the values are `HealthChecks`
- In line `//5`, with these properties, we build a `KafkaConsumer` with string keys and `HealthChecks` values, for example, `<String, HealthCheck>`

For a consumer, we provide deserializers rather than serializers. Although we don't use the key deserializer (because if you remember, it is `null`), the key deserializer is a mandatory parameter for the consumer specification. On the other hand, we need the deserializer for the value, because we are reading our data in a JSON string; here, we deserialize the object with the custom deserializer.

Java custom processor

Now, in the `src/main/java/kioto/custom` directory, create a file called `CustomProcessor.java` with the content of *Listing 4.16*.

The following is the content of *Listing 4.16*, `CustomProcessor.java` (part 1):

```
package kioto.custom;
import ...

public final class CustomProcessor {

  private Consumer<String, HealthCheck> consumer;
  private Producer<String, String> producer;

  public CustomProcessor(String brokers) {
    Properties consumerProps = new Properties();
    consumerProps.put("bootstrap.servers", brokers);
    consumerProps.put("group.id", "healthcheck-processor");
    consumerProps.put("key.deserializer", StringDeserializer.class);
    consumerProps.put("value.deserializer",
HealthCheckDeserializer.class);
    consumer = new KafkaConsumer<>(consumerProps);
    Properties producerProps = new Properties();
    producerProps.put("bootstrap.servers", brokers);
    producerProps.put("key.serializer", StringSerializer.class);
    producerProps.put("value.serializer", StringSerializer.class);
    producer = new KafkaProducer<>(producerProps);
  }
```

An analysis of the first part of the custom processor class includes the following:

- In the first part, we declare a consumer, as in *Listing 4.15*
- In the second part, we declare a producer, as in *Listing 4.13*

Now, in the `src/main/java/kioto/custom` directory, let's complete the `CustomProcessor.java` file with the content of *Listing 4.17*.

The following is the content of *Listing 4.17*, CustomProcessor.java (part 2):

```java
public final void process() {
    consumer.subscribe(Collections.singletonList(
            Constants.getHealthChecksTopic()));            //1
    while(true) {
      ConsumerRecords records = consumer.poll(Duration.ofSeconds(1L)); //2
      for(Object record : records) {                       //3
        ConsumerRecord it = (ConsumerRecord) record;
        HealthCheck healthCheck = (HealthCheck) it.value(); //4
        LocalDate startDateLocal
=healthCheck.getLastStartedAt().toInstant()
                .atZone(ZoneId.systemDefault()).toLocalDate();        //5
        int uptime =
            Period.between(startDateLocal, LocalDate.now()).getDays();
//6
        Future future =
            producer.send(new ProducerRecord<>(
                        Constants.getUptimesTopic(),
                        healthCheck.getSerialNumber(),
                        String.valueOf(uptime)));          //7
        try {
          future.get();
        } catch (InterruptedException | ExecutionException e) {
          // deal with the exception
        }
      }
    }
  }
  public static void main( String[] args) {
    new CustomProcessor("localhost:9092").process();
  }
}
```

An analysis of the CustomProcessor process method includes the following:

- In line //1, here the consumer is created and subscribed to the source topic. This is a dynamic assignment of the partitions to our customer and join to the customer group.
- In line //2, an infinite loop to consume the records, the pool duration is passed as a parameter to the method pool. The customer waits no longer than one second before return.
- In line //3, we iterate over the records.
- In line //4, the JSON string is deserialized to extract the HealthCheck object.

- In line `//5`, the start time is transformed in format at the current time zone.
- In line `//6`, the uptime is calculated.
- In line `//7`, the uptime is written to the `uptimes` topic, using the serial number as the key and the uptime as the value. Both values are written as normal strings.

Running the custom processor

To build the project, run the following command from the `kioto` directory:

```
$ gradle jar
```

If everything is correct, the output is something like the following:

```
BUILD SUCCESSFUL in 3s
1 actionable task: 1 executed
```

1. Run a console consumer for the `uptimes` topic as follows:

   ```
   $ ./bin/kafka-console-consumer --bootstrap-server localhost:9092
   --topic uptimes --property print.key=true
   ```

2. From our IDE, run the main method of `CustomProcessor`
3. From our IDE, run the main method of `CustomProducer`
4. The output on the console consumer for the `uptimes` topic should be similar to the following:

   ```
   EW05-HV36    33
   BO58-SB28    20
   DV03-ZT93    46
   . . .
   ```

Now, we have seen how to create our own SerDe to abstract the serialization code from our application's main logic. Now you know how a Kafka SerDe works.

Summary

In this chapter, we learned how to build a Java PlainProducer, a consumer, and a processor, and we have shown how to build a custom serializer and a custom deserializer.

Also, we learned how to build a Java CustomProducer, a consumer, and a processor, and how to run the Java CustomProducer and the processor.

In this chapter, we have seen how to serialize/deserialize with Kafka using JSON, plain, and binary formats. Avro is a common serialization type for Kafka. We will see how to use Avro in `Chapter 5`, *Schema Registry*, along with the use of the Kafka schema registry.

5
Schema Registry

In the previous chapter, we saw how to produce and consume data in JSON format. In this chapter, we will see how to serialize the same messages with Apache Avro.

This chapter covers the following topics:

- Avro in a nutshell
- Defining the schema
- Starting the Schema Registry
- Using the Schema Registry
- How to build a Java `AvroProducer`, a consumer, and a processor
- How to run the Java `AvroProducer` and the processor

Avro in a nutshell

Apache Avro is a binary serialization format. The format is schema-based so, it depends on the definition of schemas in JSON format. These schemas define which fields are mandatory and their types. Avro also supports arrays, enums, and nested fields.

One major advantage of Avro is that it supports schema evolution. In this way, we can have several historical versions of the schema.

Normally, the system must adapt to the changing needs of the business. For this reason, we can add or remove fields from our entities, and even change the data types. To support forward or backward compatibility, we must consider which fields are indicated as optional.

Because Avro converts the data into arrays of bytes (serialization), and Kafka's messages are also sent in binary data format, with Apache Kafka, we can send messages in Avro format. The real question is, where do we store the schemas for Apache Avro to work?

Recall that one of the main functions of an enterprise service bus is the format validation of the messages it processes, and what better if it has a historical record of these formats?

The Kafka Schema Registry is the module responsible for performing important functions. The first is to validate that the messages are in the appropriate format, the second is to have a repository of these schemas, and the third is to have a historical version format of these schemas.

The Schema Registry is a server that runs in the same place as our Kafka brokers. It runs and stores the schemas, including the schema versions. When messages are sent to Kafka in Avro format, the messages contain an identifier of a schema stored in the Schema Registry.

There is a library that allows for message serialization and deserialization in Avro format. This library works transparently and naturally with the Schema Registry.

When a message is sent in Avro format, the serializer ensures that the schema is registered and obtains the schema ID. If we send an Avro message that is not in the Schema Registry, the current version of the schema is registered automatically in the Registry. If you do not want the Schema Registry to behave in this way, you can disable it by setting the `auto.register.schemas` flag to `false`.

When a message is received in Avro format, the deserializer tries to find the schema ID in the Registry and fetch the schema to deserialize the message in Avro format.

Both the Schema Registry and the library for the serialization and deserialization of messages in Avro format are under the Confluent Platform. It is important to mention that when you need to use the Schema Registry, you must use the Confluent Platform.

It is also important to mention that with the Schema Registry, the Confluent library should be used for serialization in Avro format, as the Apache Avro library doesn't work.

Defining the schema

The first step is to define the Avro schema. As a reminder, our `HealthCheck` class looks like *Listing 5.1*:

```
public final class HealthCheck {
  private String event;
  private String factory;
  private String serialNumber;
  private String type;
  private String status;
  private Date lastStartedAt;
```

```
    private float temperature;
    private String ipAddress;
}
```

Listing 5.1: HealthCheck.java

Now, with the representation of this message in Avro format, the schema (that is, the template) of all the messages of this type in Avro would be *Listing 5.2*:

```
{
  "name": "HealthCheck",
  "namespace": "kioto.avro",
  "type": "record",
  "fields": [
  { "name": "event", "type": "string" },
  { "name": "factory", "type": "string" },
  { "name": "serialNumber", "type": "string" },
  { "name": "type", "type": "string" },
  { "name": "status", "type": "string"},
  { "name": "lastStartedAt", "type": "long", "logicalType": "timestamp-
    millis"},
  { "name": "temperature", "type": "float" },
  { "name": "ipAddress", "type": "string" }
  ]
}
```

Listing 5.2: healthcheck.avsc

This file must be saved in the `kioto` project in the `src/main/resources` directory.

It is important to note that there are the types `string`, `float`, and `double`. But, in the case of `Date`, it can be stored as a `long` or as a `string`.

For this example, we will serialize `Date` as a `long`. Avro doesn't have a dedicated `Date` type; we have to choose between a `long` and a `string` (an ISO-8601 `string` is usually better), but the point with this example is to show how to use different data types.

For more information about Avro schemas and how to map the types, check the Apache Avro specification at the following URL:
`http://avro.apache.org/docs/current/spec.html`

Starting the Schema Registry

Well, we have our Avro schema; now, we need to register it in the Schema Registry. When we start the Confluent Platform, the Schema Registry is started, as shown in the following code:

```
$./bin/confluent start
Starting zookeeper
zookeeper is [UP]
Starting kafka
kafka is [UP]
Starting schema-registry
schema-registry is [UP]
Starting kafka-rest
kafka-rest is [UP]
Starting connect
connect is [UP]
Starting ksql-server
ksql-server is [UP]
Starting control-center
control-center is [UP]
```

If we want just to start the Schema Registry, we need to run the following command:

```
$./bin/schema-registry-start etc/schema-registry/schema-registry.properties
```

The output is similar to the one shown here:

```
. . .
[2017-03-02 10:01:45,320] INFO Started
NetworkTrafficServerConnector@2ee67803{HTTP/1.1, [http/1.1]}{0.0.0.0:8081}
```

Using the Schema Registry

Now, the Schema Registry is running on port 8081. To interact with the Schema Registry, there is a REST API. We can access it with curl. The first step is to register a schema in the Schema Registry. To do so, we have to embed our JSON schema in another JSON object, and we have to escape some special characters and add a payload:

- At the beginning, we have to add { \"schema\": \"
- All the double quotation marks (") should be escaped with a backslash (\")
- At the end, we have to add \" }

Yes, as you can guess, the API has several commands to query the Schema Registry.

Registering a new version of a schema under a – value subject

To register the Avro schema `healthcheck.avsc`, located in the `src/main/resources/` path listed in *Listing 5.2*, using the `curl` command, we use the following:

```
$ curl -X POST -H "Content-Type: application/vnd.schemaregistry.v1+json" \
--data '{ "schema": "{ \"name\": \"HealthCheck\", \"namespace\":
\"kioto.avro\", \"type\": \"record\", \"fields\": [ { \"name\": \"event\",
\"type\": \"string\" }, { \"name\": \"factory\", \"type\": \"string\" }, {
\"name\": \"serialNumber\", \"type\": \"string\" }, { \"name\": \"type\",
\"type\": \"string\" }, { \"name\": \"status\", \"type\": \"string\"}, {
\"name\": \"lastStartedAt\", \"type\": \"long\", \"logicalType\":
\"timestamp-millis\"}, { \"name\": \"temperature\", \"type\": \"float\" },
{ \"name\": \"ipAddress\", \"type\": \"string\" } ]} " }' \
http://localhost:8081/subjects/healthchecks-avro-value/versions
```

The output should be something like this:

```
{"id":1}
```

This means that we have registered the `HealthChecks` schema with the version `"id":1` (congratulations, your first version).

Note that the command registers the schema on a subject called `healthchecks-avro-value`. The Schema Registry doesn't have information about topics (we still haven't created the `healthchecks-avro` topic). It is a convention, followed by the serializers/deserializers, to register schemas under a name following the <topic>-value format. In this case, since the schema is used for the message values, we use the suffix-value. If we wanted to use Avro to identify our messages keys, we would use the <topic>-key format.

For example, to obtain the ID of our schema, we use the following command:

```
$ curl http://localhost:8081/subjects/healthchecks-avro-value/versions/
```

The following output is the schema ID:

```
[1]
```

With the schema ID, to check the value of our schema, we use the following command:

```
$ curl http://localhost:8081/subjects/healthchecks-avro-value/versions/1
```

The output is the schema value shown here:

```
{"subject":"healthchecks-avro-
value","version":1,"id":1,"schema":"{\"type\":\"record\",\"name\":\"HealthC
heck\",\"namespace\":\"kioto.avro\",\"fields\":[{\"name\":\"event\",\"type\
":\"string\"},{\"name\":\"factory\",\"type\":\"string\"},{\"name\":\"serial
Number\",\"type\":\"string\"},{\"name\":\"type\",\"type\":\"string\"},{\"na
me\":\"status\",\"type\":\"string\"},{\"name\":\"lastStartedAt\",\"type\":\
"long\",\"logicalType\":\"timestamp-
millis\"},{\"name\":\"temperature\",\"type\":\"float\"},{\"name\":\"ipAddre
ss\",\"type\":\"string\"}]}"}
```

Registering a new version of a schema under a – key subject

As an example, to register a new version of our schema under the `healthchecks-avro-key` subject, we would execute the following command (don't run it; it is just to exemplify):

```
curl -X POST -H "Content-Type: application/vnd.schemaregistry.v1+json"\
--data 'our escaped avro data' \
http://localhost:8081/subjects/healthchecks-avro-key/versions
```

The output should be something like this:

```
{"id":1}
```

Registering an existing schema into a new subject

Let's suppose that there is an existing schema registered on a subject called `healthchecks-value1`, and we need this schema available on a subject called `healthchecks-value2`.

The following command reads the existing schema from `healthchecks-value1` and registers it to `healthchecks-value2` (assuming that the `jq` tool is already installed):

```
curl -X POST -H "Content-Type: application/vnd.schemaregistry.v1+json"\
--data "{\"schema\": $(curl -s
http://localhost:8081/subjects/healthchecks-value1/versions/latest | jq
'.schema')}" \
http://localhost:8081/subjects/healthchecks-value2/versions
```

The output should be something like this:

```
{"id":1}
```

Listing all subjects

To list all the subjects, you can use the following command:

```
curl -X GET http://localhost:8081/subjects
```

The output should be something like this:

```
["healthcheck-avro-value","healthchecks-avro-key"]
```

Fetching a schema by its global unique ID

To fetch a schema, you can use the following command:

```
curl -X GET http://localhost:8081/schemas/ids/1
```

The output should be something like this:

```
{"schema":"{\"type\":\"record\",\"name\":\"HealthCheck\",\"namespace\":\"ki
oto.avro\",\"fields\":[{\"name\":\"event\",\"type\":\"string\"},{\"name\":\
"factory\",\"type\":\"string\"},{\"name\":\"serialNumber\",\"type\":\"strin
g\"},{\"name\":\"type\",\"type\":\"string\"},{\"name\":\"status\",\"type\":
\"string\"},{\"name\":\"lastStartedAt\",\"type\":\"long\",\"logicalType\":\
"timestamp-
millis\"},{\"name\":\"temperature\",\"type\":\"float\"},{\"name\":\"ipAddre
ss\",\"type\":\"string\"}]}"}
```

Listing all schema versions registered under the healthchecks–value subject

To list all schema versions registered under the `healthchecks-value` subject, you can use the following command:

```
curl -X GET http://localhost:8081/subjects/healthchecks-value/versions
```

The output should be something like this:

```
[1]
```

Fetching version 1 of the schema registered under the healthchecks-value subject

To fetch version 1 of the schema registered under the `healthchecks-value` subject, you can use the following command:

```
curl -X GET http://localhost:8081/subjects/ healthchecks-value/versions/1
```

The output should be something like this:

```
{"subject":" healthchecks-value","version":1,"id":1}
```

Deleting version 1 of the schema registered under the healthchecks-value subject

To delete version 1 of the schema registered under the `healthchecks-value` subject, you can use the following command:

```
curl -X DELETE http://localhost:8081/subjects/healthchecks-value/versions/1
```

The output should be something like this:

```
1
```

Deleting the most recently registered schema under the healthchecks-value subject

To delete the most recently registered schema under the `healthchecks-value` subject, you can use the following command:

```
curl -X DELETE
http://localhost:8081/subjects/healthchecks-value/versions/latest
```

The output should be something like this:

```
2
```

Deleting all the schema versions registered under the healthchecks–value subject

To delete all the schema versions registered under the `healthchecks-value` subject, you can use the following command:

```
curl -X DELETE http://localhost:8081/subjects/healthchecks-value
```

The output should be something like this:

```
[3]
```

Checking whether a schema is already registered under the healthchecks–key subject

To check whether a schema is already registered under the `healthchecks-key` subject, you can use the following command:

```
curl -X POST -H "Content-Type: application/vnd.schemaregistry.v1+json"\
--data 'our escaped avro data' \
http://localhost:8081/subjects/healthchecks-key
```

The output should be something like this:

```
{"subject":"healthchecks-key","version":3,"id":1}
```

Testing schema compatibility against the latest schema under the healthchecks–value subject

To test the schema compatibility against the latest schema under the `healthchecks-value` subject, you can use the following command:

```
curl -X POST -H "Content-Type: application/vnd.schemaregistry.v1+json"\
--data 'our escaped avro data' \
http://localhost:8081/compatibility/subjects/healthchecks-value/versions/la
test
```

The output should be something like this:

```
{"is_compatible":true}
```

Getting the top-level compatibility configuration

To get the top level compatibility configuration, you can use the following command:

```
curl -X GET http://localhost:8081/config
```

The output should be something like this:

```
{"compatibilityLevel":"BACKWARD"}
```

Globally updating the compatibility requirements

To globally update the compatibility requirements, you can use the following command:

```
curl -X PUT -H "Content-Type: application/vnd.schemaregistry.v1+json" \
--data '{"compatibility": "NONE"}' \
http://localhost:8081/config
```

The output should be something like this:

```
{"compatibility":"NONE"}
```

Updating the compatibility requirements under the healthchecks–value subject

To update the compatibility requirements under the `healthchecks-value` subject, you can use the following command:

```
curl -X PUT -H "Content-Type: application/vnd.schemaregistry.v1+json" \
--data '{"compatibility": "BACKWARD"}' \
http://localhost:8081/config/healthchecks-value
```

The output should be something like this:

```
{"compatibility":"BACKWARD"}
```

Java AvroProducer

Now, we should modify our Java Producer to send messages in Avro format. First, it is important to mention that in Avro there are two types of messages:

- **Specific records**: The file with the Avro schema (avsc) is sent to a specific Avro command to generate the corresponding Java classes.
- **Generic records**: In this approach, a data structure similar to a map dictionary is used. This means that you set and get the fields by their names and you must know their corresponding types. This option is not type-safe, but it offers much more flexibility than the other, and here the versions are much easier to manage over time. In this example, we will use this approach.

Before we start with the code, remember that in the last chapter we added the library to support Avro to our Kafka client. If you recall, the `build.gradle` file has a special repository with all this libraries.

Confluent's repository is specified in the following line:

```
repositories {
  ...
  maven { url 'https://packages.confluent.io/maven/' }
}
```

In the dependencies section, we should add the specific Avro libraries:

```
dependencies {
  ...
  compile 'io.confluent:kafka-avro-serializer:5.0.0'
}
```

Do not use the libraries provided by Apache Avro, because they will not work.
As we already know, to build a Kafka message producer, we use the Java client library; in particular, the producer API. As we already know, there are two requisites that all the Kafka producers should have: to be a `KafkaProducer` and to set the specific `Properties`, such as *Listing 5.3*:

```
import io.confluent.kafka.serializers.KafkaAvroSerializer;
import org.apache.avro.Schema;
import org.apache.avro.Schema.Parser;
import org.apache.avro.generic.GenericRecord;
import org.apache.kafka.clients.producer.KafkaProducer;
import org.apache.kafka.clients.producer.Producer;
import org.apache.kafka.common.serialization.StringSerializer;
```

```
public final class AvroProducer {
  private final Producer<String, GenericRecord> producer; //1
  private Schema schema;
  public AvroProducer(String brokers, String schemaRegistryUrl) { //2
    Properties props = new Properties();
    props.put("bootstrap.servers", brokers);
    props.put("key.serializer", StringSerializer.class); //3
    props.put("value.serializer", KafkaAvroSerializer.class); //4
    props.put("schema.registry.url", schemaRegistryUrl) //5
    producer = new KafkaProducer<>(props);

    try {
      schema = (new Parser()).parse( new
      File("src/main/resources/healthcheck.avsc")); //6
    } catch (IOException e) {
      // deal with the Exception
    }
  }
  ...
}
```

Listing 5.3: AvroProducer Constructor

An analysis of the AvroProducer constructor shows the following:

- In line //1, the values now are of type org.apache.avro.generic.GenericRecord
- In line //2, the constructor now receives the Schema Registry URL
- In line //3, the Serializer type for the messages' keys remains as StringSerializer
- In line //4, the Serializer type for the messages' values now is a KafkaAvroSerializer
- In line //5, the Schema Registry URL is added to the Producer properties
- In line //6, the avsc file with the schema definition is parsed with a Schema Parser

Because we have chosen the use of generic records, we have to load the schema. Note that we could have obtained the schema from the Schema Registry, but this is not safe because we do not know which version of the schema is registered. Instead of this, it is a smart and safe practice to store the schema along with the code. In this way, our code will always produce the correct data types, even when someone else changes the schema registered in the Schema Registry.

Now, in the `src/main/java/kioto/avro` directory, create a file called `AvroProducer.java` with the contents of *Listing 5.4*:

```
package kioto.avro;
import ...
public final class AvroProducer {
 /* here the Constructor code in Listing 5.3 */

  public final class AvroProducer {

    private final Producer<String, GenericRecord> producer;
    private Schema schema;

    public AvroProducer(String brokers, String schemaRegistryUrl) {
      Properties props = new Properties();
      props.put("bootstrap.servers", brokers);
      props.put("key.serializer", StringSerializer.class);
      props.put("value.serializer", KafkaAvroSerializer.class);
      props.put("schema.registry.url", schemaRegistryUrl);
      producer = new KafkaProducer<>(props);
      try {
        schema = (new Parser()).parse(new
        File("src/main/resources/healthcheck.avsc"));
      } catch (IOException e) {
        e.printStackTrace();
      }
    }

    public final void produce(int ratePerSecond) {
      long waitTimeBetweenIterationsMs = 1000L / (long)ratePerSecond;
      Faker faker = new Faker();

      while(true) {
        HealthCheck fakeHealthCheck =
            new HealthCheck(
                "HEALTH_CHECK",
                faker.address().city(),
                faker.bothify("??##-??##", true),
                Constants.machineType.values()
                [faker.number().numberBetween(0,4)].toString(),
                Constants.machineStatus.values()
                [faker.number().numberBetween(0,3)].toString(),
                faker.date().past(100, TimeUnit.DAYS),
                faker.number().numberBetween(100L, 0L),
                faker.internet().ipV4Address());
                GenericRecordBuilder recordBuilder = new
                GenericRecordBuilder(schema);
                recordBuilder.set("event", fakeHealthCheck.getEvent());
```

```
                    recordBuilder.set("factory",
                    fakeHealthCheck.getFactory());
                    recordBuilder.set("serialNumber",
                    fakeHealthCheck.getSerialNumber());
                    recordBuilder.set("type", fakeHealthCheck.getType());
                    recordBuilder.set("status", fakeHealthCheck.getStatus());
                    recordBuilder.set("lastStartedAt",
                    fakeHealthCheck.getLastStartedAt().getTime());
                    recordBuilder.set("temperature",
                    fakeHealthCheck.getTemperature());
                    recordBuilder.set("ipAddress",
                    fakeHealthCheck.getIpAddress());
                    Record avroHealthCheck = recordBuilder.build();
                    Future futureResult = producer.send(new ProducerRecord<>
                    (Constants.getHealthChecksAvroTopic(), avroHealthCheck));
            try {
              Thread.sleep(waitTimeBetweenIterationsMs);
              futureResult.get();
            } catch (InterruptedException | ExecutionException e) {
              e.printStackTrace();
            }
          }
      }

    public static void main( String[] args) {
      new AvroProducer("localhost:9092",
      "http://localhost:8081").produce(2);
    }
  }
```

Listing 5.4: AvroProducer.java

An analysis of the AvroProducer class shows the following:

- In line //1, ratePerSecond is the number of messages to send in a 1-second period
- In line //2, to simulate repetition, we use an infinite loop (try to avoid this in production)
- In line //3, now we can create GenericRecord objects using GenericRecordBuilder
- In line //4, we use a Java Future to send the record to the healthchecks-avro topic

- In line //5, we wait this time to send messages again
- In line //6, we read the result of the Future
- In line //7, everything runs on the broker on the localhost in port 9092, and with the Schema Registry running on the localhost in port 8081, sending two messages in an interval of 1 second

Running the AvroProducer

To build the project, run the following command from the `kioto` directory:

```
$ gradle jar
```

If everything is OK, the output is something like the one shown here:

```
BUILD SUCCESSFUL in 3s
1 actionable task: 1 executed
```

1. If it is not running yet, go to Confluent's directory and start it:

```
$ ./bin/confluent start
```

2. The broker is running on port 9092. To create the `healthchecks-avro` topic, execute the following command:

```
$ ./bin/kafka-topics --zookeeper localhost:2181 --create --topic
healthchecks-avro --replication-factor 1 --partitions 4
```

3. Note that we are just creating a normal topic and nothing indicates the messages' format.

4. Run a console consumer for the `healthchecks-avro` topic:

```
$ ./bin/kafka-console-consumer --bootstrap-server localhost:9092 --
topic healthchecks-avro
```

5. From our IDE, run the main method of the `AvroProducer`.

6. The output on the console consumer should be similar to the one shown here:

```
HEALTH_CHECKLake JeromyGE50-
GF78HYDROELECTRICRUNNING◆◆◆◆◆Y,B227.30.250.185
HEALTH_CHECKLockmanlandMW69-LS32GEOTHERMALRUNNING◆◆◆YB72.194.121.48
HEALTH_CHECKEast IsidrofortIH27-
WB64NUCLEARSHUTTING_DOWN◆◆◆YB88.136.134.241
HEALTH_CHECKSipesshireDH05-YR95HYDROELECTRICRUNNING◆◆◆◆◆Y◆B254.125.63.235
HEALTH_CHECKPort EmeliaportDJ83-UO93GEOTHERMALRUNNING◆◆◆Y◆A190.160.48.125
```

Binary is a horrible format for humans to read, isn't it? We can just read the strings but not the rest of the record.

To solve our readability problem, we should use `kafka-avro-console-consumer` instead. This fancy consumer deserializes the Avro records and prints them as human-readable JSON objects.

From the command line, run an Avro console consumer for the `healthchecks-avro` topic:

```
$ ./bin/kafka-avro-console-consumer --bootstrap-server localhost:9092 --
topic healthchecks-avro
```

The output on the console consumer should be similar to this:

```
{"event":"HEALTH_CHECK","factory":"Lake Jeromy","serialNumber":" GE50-
GF78","type":"HYDROELECTRIC","status":"RUNNING","lastStartedAt":15373207199
54,"temperature":35.0,"ipAddress":"227.30.250.185"}
{"event":"HEALTH_CHECK","factory":"Lockmanland","serialNumber":" MW69-
LS32","type":"GEOTHERMAL","status":"RUNNING","lastStartedAt":1534188452893,
"temperature":61.0,"ipAddress":"72.194.121.48"}
{"event":"HEALTH_CHECK","factory":"East Isidrofort","serialNumber":" IH27-
WB64","type":"NUCLEAR","status":"SHUTTING_DOWN","lastStartedAt":15392964031
79,"temperature":62.0,"ipAddress":"88.136.134.241"}
. . .
```

Now, we are finally producing Kafka messages in Avro format. With the help of the Schema Registry and the Confluent library, this task is quite simple. As described, after much frustration in productive environments, the generic records scheme is better than the specific records scheme, because it is better to know specifically with which schema we are producing data. Keeping a copy of the schema along with the code gives you that guarantee.

What happens if we fetch the schema from the Schema Registry before producing the data? The correct answer is it depends, and it depends on the `auto.register.schemas` property. If this property is set to true, when you request a schema that is not in the Schema Registry, it will automatically be registered as a new schema (this option in production environments is not recommended, as it is prone to error). If the property is set to false, the schema will not be stored and, since the schema will not match, we will have a nice exception (do not believe me, reader; go and get proof of this).

Java AvroConsumer

Let's create a Kafka `AvroConsumer` that we will use to receive the input records. As we already know, there are two prerequisites that all the Kafka Consumers should have: to be a `KafkaConsumer` and to set specific properties, such as in *Listing 5.5*:

```
import io.confluent.kafka.serializers.KafkaAvroDeserializer;
import org.apache.avro.generic.GenericRecord;
import org.apache.kafka.clients.consumer.KafkaConsumer;
import org.apache.kafka.clients.consumer.Consumer;
import org.apache.kafka.common.serialization.StringSerializer;

public final class AvroConsumer {
  private Consumer<String, GenericRecord> consumer; //1
  public AvroConsumer(String brokers, String schemaRegistryUrl) { //2
    Properties props = new Properties();
    props.put("group.id", "healthcheck-processor");
    props.put("bootstrap.servers", brokers);
    props.put("key.deserializer", StringDeserializer.class); //3
    props.put("value.deserializer", KafkaAvroDeserializer.class); //4
    props.put("schema.registry.url", schemaRegistryUrl); //5
    consumer = new KafkaConsumer<>(props); //6
  }
  ...
}
```

Listing 5.5: AvroConsumer constructor

An analysis of the changes in the `AvroConsumer` constructor shows the following:

- In line `//1`, the values now are of type `org.apache.avro.generic.GenericRecord`
- In line `//2`, the constructor now receives the Schema Registry URL
- In line `//3`, the deserializer type for the messages' keys remains as `StringDeserializer`
- In line `//4`, the deserializer type for the values is now `KafkaAvroDeserializer`
- In line `//5`, the Schema Registry URL is added to the consumer properties
- In line `//6`, with these `Properties`, we build a `KafkaConsumer` with string keys and `GenericRecord` values: `<String, GenericRecord>`

It is important to note that when defining the Schema Registry URL for the deserializer to fetch schemas, the messages only contain the schema ID and not the schema itself.

Java AvroProcessor

Now, in the `src/main/java/kioto/avro` directory, create a file called `AvroProcessor.java` with the contents of *Listing 5.6*:

```
package kioto.plain;
import ...
public final class AvroProcessor {
  private Consumer<String, GenericRecord> consumer;
  private Producer<String, String> producer;

  public AvroProcessor(String brokers , String schemaRegistryUrl) {
    Properties consumerProps = new Properties();
    consumerProps.put("bootstrap.servers", brokers);
    consumerProps.put("group.id", "healthcheck-processor");
    consumerProps.put("key.deserializer", StringDeserializer.class);
    consumerProps.put("value.deserializer", KafkaAvroDeserializer.class);
    consumerProps.put("schema.registry.url", schemaRegistryUrl);
    consumer = new KafkaConsumer<>(consumerProps);
    Properties producerProps = new Properties();
    producerProps.put("bootstrap.servers", brokers);
    producerProps.put("key.serializer", StringSerializer.class);
    producerProps.put("value.serializer", StringSerializer.class);
    producer = new KafkaProducer<>(producerProps);
  }
```

Listing 5.6: AvroProcessor.java (part 1)

An analysis of the first part of the `AvroProcessor` class shows the following:

- In the first section, we declare an `AvroConsumer`, as in *Listing 5.5*
- In the second section, we declare an `AvroProducer`, as in *Listing 5.4*

Now, in the `src/main/java/kioto/avro` directory, let's complete the `AvroProcessor.java` file with the contents of *Listing 5.7*:

```
public final void process() {
  consumer.subscribe(Collections.singletonList(
    Constants.getHealthChecksAvroTopic())); //1
    while(true) {
      ConsumerRecords records = consumer.poll(Duration.ofSeconds(1L));
      for(Object record : records) {
        ConsumerRecord it = (ConsumerRecord) record;
        GenericRecord healthCheckAvro = (GenericRecord) it.value(); //2
        HealthCheck healthCheck = new HealthCheck ( //3
          healthCheckAvro.get("event").toString(),
```

```
            healthCheckAvro.get("factory").toString(),
            healthCheckAvro.get("serialNumber").toString(),
            healthCheckAvro.get("type").toString(),
            healthCheckAvro.get("status").toString(),
            new Date((Long)healthCheckAvro.get("lastStartedAt")),
            Float.parseFloat(healthCheckAvro.get("temperature").toString()),
            healthCheckAvro.get("ipAddress").toString());
            LocalDate startDateLocal=
            healthCheck.getLastStartedAt().toInstant()
                    .atZone(ZoneId.systemDefault()).toLocalDate(); //4
            int uptime = Period.between(startDateLocal,
            LocalDate.now()).getDays(); //5
            Future future =
                producer.send(new ProducerRecord<>(
                            Constants.getUptimesTopic(),
                            healthCheck.getSerialNumber(),
                            String.valueOf(uptime))); //6
            try {
              future.get();
            } catch (InterruptedException | ExecutionException e) {
              // deal with the exception
            }
          }
        }
    }

    public static void main(String[] args) {
        new
  AvroProcessor("localhost:9092","http://localhost:8081").process();//7
    }
}
```

Listing 5.7: AvroProcessor.java (part 2)

An analysis of the `AvroProcessor` shows the following:

- In line `//1`, the consumer is subscribed to the new Avro topic.
- In line `//2`, we are consuming messages of type `GenericRecord`.
- In line `//3`, the Avro record is deserialized to extract the `HealthCheck` object.
- In line `//4`, the start time is transformed into the format in the current time zone.
- In line `//5`, the uptime is calculated.
- In line `//6`, the uptime is written to the `uptimes` topic, using the serial number as the key and the uptime as the value. Both values are written as normal strings.
- In line `//7`, everything runs on the broker on the localhost in port `9092` and with the Schema Registry running on the localhost in port `8081`.

As mentioned previously, the code is not type-safe; all the types are checked at runtime. So, be extremely careful with that. For example, the strings are not `java.lang.String`; they are of type `org.apache.avro.util.Utf8`. Note that we avoid the cast by calling the `toString()` method directly on the objects. The rest of the code remains equal.

Running the AvroProcessor

To build the project, run the following command from the `kioto` directory:

```
$ gradle jar
```

If everything is correct, the output will be something like this:

```
BUILD SUCCESSFUL in 3s
  1 actionable task: 1 executed
```

Run a console consumer for the `uptimes` topic, as shown here:

```
$ ./bin/kafka-console-consumer --bootstrap-server localhost:9092 --topic
uptimes --property print.key=true
```

1. From the IDE, run the main method of the `AvroProcessor`
2. From the IDE, run the main method of the `AvroProducer`
3. The output on the console consumer for the `uptimes` topic should be similar to this:

```
EW05-HV36 33
BO58-SB28 20
DV03-ZT93 46
. . .
```

Summary

In this chapter, we showed, instead of sending data in JSON format, how to use AVRO as the serialization format. The main benefit of AVRO (over JSON, for example) is that the data must conform to the schema. Another advantage of AVRO over JSON is that the messages are more compact when sent in binary format, although JSON is human readable.

The schemas are stored in the Schema Registry, so that all users can consult the schema version history, even when the code of the producers and consumers for those messages is no longer available.

Apache Avro also guarantees backward and forward compatibility of all messages in this format. Forward compatibility is achieved following some basic rules, such as when adding a new field, declaring its value as optional.

Apache Kafka encourages the use of Apache Avro and the Schema Registry for the storage of all data and schemas in Kafka systems, instead of using only plain text or JSON. Using that winning combination, you guarantee that your system can evolve.

6
Kafka Streams

In this chapter, instead of using the Kafka Java API for producers and consumers as in previous chapters, we are going to use Kafka Streams, the Kafka module for stream processing.

This chapter covers the following topics:

- Kafka Streams in a nutshell
- Kafka Streams project setup
- Coding and running the Java `PlainStreamsProcessor`
- Scaling out with Kafka Streams
- Coding and running the Java `CustomStreamsProcessor`
- Coding and running the Java `AvroStreamsProcessor`
- Coding and running the Late `EventProducer`
- Coding and running the Kafka Streams processor

Kafka Streams in a nutshell

Kafka Streams is a library and part of Apache Kafka, used to process streams into and from Kafka. In functional programming, there are several operations over collections, such as the following:

- `filter`
- `map`
- `flatMap`
- `groupBy`
- `join`

The success of streaming platforms such as Apache Spark, Apache Flink, Apache Storm, and Akka Streams is to incorporate these stateless functions to process data streams. Kafka Streams provides a DSL to incorporate these functions to manipulate data streams. Kafka Streams also has stateful transformations; these are operations related to the aggregation that depend on the state of the messages as a group, for example, the windowing functions and support for late arrival data. Kafka Streams is a library, and this means that Kafka Streams applications can be deployed by executing your application jar. There is no need to deploy the application on a server, which means you can use any application to run a Kafka Streams application: Docker, Kubernetes, servers on premises, and so on. Something wonderful about Kafka Streams is that it allows horizontal scaling. That is, if it runs in the same JVM, it executes multiple threads, but if several instances of the application are started, it can run several JVMs to scale out.

The Apache Kafka core is built in Scala; however, Kafka Streams and KSQL are built in Java 8. Kafka Streams is packaged in the open source distribution of Apache Kafka.

Project setup

The first step is to modify the `kioto` project. We have to add the dependencies to `build.gradle`, as shown in *Listing 6.1*:

```
apply plugin: 'java'
apply plugin: 'application'

sourceCompatibility = '1.8'

mainClassName = 'kioto.ProcessingEngine'

repositories {
 mavenCentral()
 maven { url 'https://packages.confluent.io/maven/' }
}

version = '0.1.0'

dependencies {
   compile 'com.github.javafaker:javafaker:0.15'
   compile 'com.fasterxml.jackson.core:jackson-core:2.9.7'
   compile 'io.confluent:kafka-avro-serializer:5.0.0'
   compile 'org.apache.kafka:kafka_2.12:2.0.0'
   compile 'org.apache.kafka:kafka-streams:2.0.0'
   compile 'io.confluent:kafka-streams-avro-serde:5.0.0'
}
```

```
jar {
  manifest {
    attributes 'Main-Class': mainClassName
  } from {
    configurations.compile.collect {
        it.isDirectory() ? it : zipTree(it)
    }
  }
  exclude "META-INF/*.SF"
  exclude "META-INF/*.DSA"
  exclude "META-INF/*.RSA"
}
```

Listing 6.1: Kioto Gradle build file for Kafka Streams

For the examples in this chapter, we also need the dependencies for Jackson. To use Kafka Streams, we just need one dependency, which is given in the following code snippet:

```
compile 'org.apache.kafka:kafka-streams:2.0.0'
```

To use Apache Avro with Kafka Streams, we add the serializers and deserializers as given in the following code:

```
compile 'io.confluent:kafka-streams-avro-serde:5.0.0'
```

The following lines are needed to run a Kafka Streams application as a jar. The build generates a fat jar:

```
configurations.compile.collect {
  it.isDirectory() ? it : zipTree(it)
}
```

The directory tree structure of the project should be as follows:

```
src
main
--java
----kioto
------avro
------custom
------events
------plain
------serde
--resources
test
```

Java PlainStreamsProcessor

Now, in the `src/main/java/kioto/plain` directory, create a file called `PlainStreamsProcessor.java` with the contents of *Listing 6.2,* shown as follows:

```
import ...
public final class PlainStreamsProcessor {
  private final String brokers;
  public PlainStreamsProcessor(String brokers) {
    super();
    this.brokers = brokers;
  }
  public final void process() {
    // below we will see the contents of this method
  }
  public static void main(String[] args) {
    (new PlainStreamsProcessor("localhost:9092")).process();
  }
}
```

Listing 6.2: PlainStreamsProcessor.java

All the magic happens inside the `process()` method. The first step in a Kafka Streams application is to get a `StreamsBuilder` instance, as shown in the following code:

```
StreamsBuilder streamsBuilder = new StreamsBuilder();
```

The `StreamsBuilder` is an object that allows building a topology. A topology in Kafka Streams is a structural description of a data pipeline. The topology is a succession of steps that involve transformations between streams. A topology is a very important concept in streams; it is also used in other technologies such as Apache Storm.

The `StreamsBuilder` is used to consume data from a topic. There are other two important concepts in the context of Kafka Streams: a `KStream`, a representation of a stream of records, and a `KTable`, a log of the changes in a stream (we will see KTables in detail in Chapter 7, *KSQL*). To obtain a `KStream` from a topic, we use the `stream()` method of the `StreamsBuilder`, shown as follows:

```
KStream healthCheckJsonStream =
  streamsBuilder.stream( Constants.getHealthChecksTopic(),
    Consumed.with(Serdes.String(), Serdes.String()));
```

There is an implementation of the `stream()` method that just receives the topic name as a parameter. But, it is good practice to use the implementation where we can also specify the serializers, as in this example we have to specify the Serializer for the key and the Serializer for the value for the `Consumed` class; in this case, both are strings.

Don't let the serializers be specified through application-wide properties, because the same Kafka Streams application might read from several data sources with different data formats.

We have obtained a JSON stream. The next step in the topology is to obtain the `HealthCheck` object stream, and we do so by building the following Stream:

```
KStream healthCheckStream = healthCheckJsonStream.mapValues((v -> {
  try {
    return Constants.getJsonMapper().readValue(
      (String) v, HealthCheck.class);
  } catch (IOException e) {
    // deal with the Exception
  }
}));
```

First, note that we are using the `mapValues()` method, so as in Java 8, the method receives a lambda expression. There are other implementations for the `mapValues()` method, but here we are using the lambda with just one argument (v->).

The `mapValues()` here could be read as follows: for each element in the input Stream, we are applying a transformation from the JSON object to the `HealthCheck` object, and this transformation could raise an `IOException`, so we are catching it.

Recapitulating until the moment, in the first transformation, we read from the topic a stream with (`String, String`) pairs. In the second transformation, we go from the value in JSON to the value in `HealthCheck` objects.

In the third step, we are going to calculate the `uptime` and send it to the `uptimeStream`, as shown in the following block:

```
KStream uptimeStream = healthCheckStream.map(((KeyValueMapper)(k, v)-> {
  HealthCheck healthCheck = (HealthCheck) v;
  LocalDate startDateLocal = healthCheck.getLastStartedAt().toInstant()
            .atZone(ZoneId.systemDefault()).toLocalDate();
  int uptime = Period.between(startDateLocal, LocalDate.now()).getDays();
  return new KeyValue<>(
    healthCheck.getSerialNumber(), String.valueOf(uptime));
}));
```

Note that we are using the `map()` method, also as in Java 8, the method receives a lambda expression. There are other implementations for the `map()` method; here, we are using a lambda with two arguments (`(k, v)->`)

The `map()` here could be read as follows: for each element in the input stream, we extract the tuples (key, value). We are using just the value (anyway, the key is `null`), cast it to `HealthCheck`, extract two attributes (the start time and the `SerialNumber`), calculate the `uptime`, and return a new `KeyValue` pair with (`SerialNumber`, `uptime`).

The last step is to write these values into the `uptimes` topic, shown as follows:

```
uptimeStream.to( Constants.getUptimesTopic(),
    Produced.with(Serdes.String(), Serdes.String()));
```

Again, I will emphasize it until I get tired: it is widely recommended to declare the data types of our Streams. Always stating, in this case for example, that key value pairs are of type (`String`, `String`).

Here is a summary of the steps:

1. Read from the input topic key value pairs of type (`String`, `String`)
2. Deserialize each JSON object to `HealthCheck`
3. Calculate the `uptimes`
4. Write the `uptimes` to the output topic in key value pairs of type (`String`, `String`)

Finally, it is time to start the Kafka Streams engine.

Before starting it, we need to specify the topology and two properties, the broker and the application ID, shown as follows:

```
Topology topology = streamsBuilder.build();
Properties props = new Properties();
props.put("bootstrap.servers", this.brokers);
props.put("application.id", "kioto");
KafkaStreams streams = new KafkaStreams(topology, props);
streams.start();
```

Note that the serializers and deserializers are just explicitly defined when reading from and writing to topics. So, we are not tied application-wide to a single data type, and we can read from and write to topics with different data types, as happens continuously in practice.

Also with this good practice, between different topics, there is no ambiguity about which Serde to use.

Running the PlainStreamsProcessor

To build the project, run this command from the `kioto` directory:

```
$ gradle build
```

If everything is correct, the output is something like the following:

```
BUILD SUCCESSFUL in 1s
6 actionable task: 6 up-to-date
```

1. The first step is to run a console consumer for the `uptimes` topic, shown in the following code snippet:

   ```
   $ ./bin/kafka-console-consumer --bootstrap-server localhost:9092
   --topic uptimes --property print.key=true
   ```

2. From the IDE, run the main method of the `PlainStreamsProcessor`

3. From the IDE, run the main method of the `PlainProducer` (built in previous chapters)

4. The output on the console consumer for the `uptimes` topic should be similar to the following:

   ```
   EW05-HV36 33
   BO58-SB28 20
   DV03-ZT93 46
   . . .
   ```

Scaling out with Kafka Streams

To scale out the architecture as promised, we must follow these steps:

1. Run a console consumer for the `uptimes` topic, shown as follows:

   ```
   $ ./bin/kafka-console-consumer --bootstrap-server localhost:9092
   --topic uptimes --property print.key=true
   ```

2. Run the application jar from the command line, shown in the following code:

   ```
   $ java -cp ./build/libs/kioto-0.1.0.jar
   kioto.plain.PlainStreamsProcessor
   ```

   This is when we verify that our application really scales out.

3. From a new command-line window, we execute the same command, shown as follows:

```
$ java -cp ./build/libs/kioto-0.1.0.jar
kioto.plain.PlainStreamsProcessor
```

The output should be something like the following:

```
2017/07/05 15:03:18.045 INFO ... Setting newly assigned
partitions [healthchecks-2, healthchecks -3]
```

If we remember the theory of `Chapter 1`, *Configuring Kafka,* when we created our topic, we specified that it had four partitions. This nice message from Kafka Streams is telling us that the application was assigned to partitions two and three of our topic.

Take a look at the following log:

```
...
2017/07/05 15:03:18.045 INFO ... Revoking previously assigned partitions
[healthchecks -0, healthchecks -1, healthchecks -2, healthchecks -3]
2017/07/05 15:03:18.044 INFO ... State transition from RUNNING to
PARTITIONS_REVOKED
2017/07/05 15:03:18.044 INFO ... State transition from RUNNING to
REBALANCING
2017/07/05 15:03:18.044 INFO ... Setting newly assigned partitions
[healthchecks-2, healthchecks -3]
...
```

We can read that the first instance was using the four partitions, then when we ran the second instance, it entered a state where the partitions were reassigned to consumers; to the first instance was assigned two partitions: `healthchecks-0` and `healthchecks-1`.

And this is how Kafka Streams smoothly scale out. Remember that all this works because the consumers are part of the same consumer group and are controlled from Kafka Streams through the `application.id` property.

We must also remember that the number of threads assigned to each instance of our application can also be modified by setting the `num.stream.threads` property. Thus, each thread would be independent, with its own producer and consumer. This ensures that the resources of our servers are used in a more efficient way.

Java CustomStreamsProcessor

Summing up what has happened so far, in previous chapters we saw how to make a producer, a consumer, and a simple processor in Kafka. We also saw how to do the same with a custom SerDe, how to use Avro, and the Schema Registry. So far in this chapter, we have seen how to make a simple processor with Kafka Streams.

In this section, we will use all our knowledge so far to build a `CustomStreamsProcessor` with Kafka Streams to use our own SerDe.

Now, in the `src/main/java/kioto/custom` directory, create a file called `CustomStreamsProcessor.java` with the contents of *Listing 6.3*, shown as follows:

```
import ...
public final class CustomStreamsProcessor {
  private final String brokers;
  public CustomStreamsProcessor(String brokers) {
    super();
    this.brokers = brokers;
  }
  public final void process() {
    // below we will see the contents of this method
  }
  public static void main(String[] args) {
    (new CustomStreamsProcessor("localhost:9092")).process();
  }
}
```

Listing 6.3: CustomStreamsProcessor.java

All the magic happens inside the `process()` method.

The first step in a Kafka Streams application is to get a `StreamsBuilder` instance, shown as follows:

```
StreamsBuilder streamsBuilder = new StreamsBuilder();
```

We can reuse the `Serdes` built in the previous chapters. The following code creates a `KStream` that deserializes the values of the messages as `HealthCheck` objects.

```
Serde customSerde = Serdes.serdeFrom(
  new HealthCheckSerializer(), new HealthCheckDeserializer());
```

The `serdeFrom()` method of the `Serde` class dynamically wraps our `HealthCheckSerializer` and `HealthCheckDeserializer` into a single `HealthCheck` `Serde`.

We can reuse the `Serdes` built on the previous chapters. The following code creates a `KStream` that deserializes the values of the messages as `HealthCheck` objects.

The `StreamsBuilder` is used to consume data from a topic. The same as in previous sections, to obtain a `KStream` from a topic, we use the `stream()` method of the `StreamsBuilder`, shown as follows:

```
KStream healthCheckStream =
  streamsBuilder.stream( Constants.getHealthChecksTopic(),
    Consumed.with(Serdes.String(), customSerde));
```

We use the implementation where we can also specify the serializers, as in this example, we have to specify the serializer for the key, and the serializer for the value for the `Consumed` class, in this case the key is a String (always `null`), and the serializer for the value is our new `customSerde`.

The magic here is that the rest of the code of the `process()` method remains the same as in the previous section; it is also shown as follows:

```
KStream uptimeStream = healthCheckStream.map(((KeyValueMapper)(k, v)-> {
  HealthCheck healthCheck = (HealthCheck) v;
  LocalDate startDateLocal = healthCheck.getLastStartedAt().toInstant()
            .atZone(ZoneId.systemDefault()).toLocalDate();
  int uptime =
      Period.between(startDateLocal, LocalDate.now()).getDays();
  return new KeyValue<>(
      healthCheck.getSerialNumber(), String.valueOf(uptime));
}));
uptimeStream.to( Constants.getUptimesTopic(),
      Produced.with(Serdes.String(), Serdes.String()));
Topology topology = streamsBuilder.build();
Properties props = new Properties();
props.put("bootstrap.servers", this.brokers);
props.put("application.id", "kioto");
KafkaStreams streams = new KafkaStreams(topology, props);
streams.start();
```

Running the CustomStreamsProcessor

To build the project, run this command from the `kioto` directory:

```
$ gradle build
```

If everything is correct, the output is something like the following:

```
BUILD SUCCESSFUL in 1s
6 actionable task: 6 up-to-date
```

1. The first step is to run a console consumer for the `uptimes` topic, shown as follows:

   ```
   $ ./bin/kafka-console-consumer --bootstrap-server localhost:9092
   --topic uptimes --property print.key=true
   ```

2. From our IDE, run the main method of the `CustomStreamsProcessor`
3. From our IDE, run the main method of the `CustomProducer` (built in previous chapters)
4. The output on the console consumer for the `uptimes` topic should be similar to the following:

   ```
   EW05-HV36 33
   BO58-SB28 20
   DV03-ZT93 46
   . . .
   ```

Java AvroStreamsProcessor

In this section we will see how to use all this power gathered together: Apache Avro, Schema Registry, and Kafka Streams.

Now, we are going to use Avro format in our messages, as we did in previous chapters. We consumed this data by configuring the Schema Registry URL and using the Kafka Avro deserializer. For Kafka Streams, we need to use a Serde, so we added the dependency in the Gradle build file, given as follows:

```
compile 'io.confluent:kafka-streams-avro-serde:5.0.0'
```

This dependency has the `GenericAvroSerde` and specific `avroSerde` explained in previous chapters. Both Serde implementations allow us to work with Avro records.

Now, in the `src/main/java/kioto/avro` directory, create a file called `AvroStreamsProcessor.java` with the contents of *Listing 6.4,* shown as follows:

```
import ...
public final class AvroStreamsProcessor {
  private final String brokers;
  private final String schemaRegistryUrl;
  public AvroStreamsProcessor(String brokers, String schemaRegistryUrl) {
    super();
    this.brokers = brokers;
    this.schemaRegistryUrl = schemaRegistryUrl;
  }
  public final void process() {
    // below we will see the contents of this method
  }
  public static void main(String[] args) {
    (new AvroStreamsProcessor("localhost:9092",
        "http://localhost:8081")).process();
  }
}
```

Listing 6.4: AvroStreamsProcessor.java

One main difference with the previous code listings is the specification of the Schema Registry URL. The same as before, the magic happens inside the `process()` method.

The first step in a Kafka Streams Application is to get a `StreamsBuilder` instance, given as follows:

```
StreamsBuilder streamsBuilder = new StreamsBuilder();
```

The seconds step is to get an instance of the `GenericAvroSerde` object, shown as follows:

```
GenericAvroSerde avroSerde = new GenericAvroSerde();
```

As we are using the `GenericAvroSerde`, we need to configure it with the Schema Registry URL (as in previous chapters); it is shown in the following code:

```
avroSerde.configure(
  Collections.singletonMap("schema.registry.url", schemaRegistryUrl),
  false);
```

The `configure()` method of `GenericAvroSerde` receives a map as a parameter; as we just need a map with a single entry, we used the singleton map method.

Now, we can create a KStream using this Serde. The following code generates an Avro Stream that contains GenericRecord objects:

```
KStream avroStream =
  streamsBuilder.stream( Constants.getHealthChecksAvroTopic(),
    Consumed.with(Serdes.String(), avroSerde));
```

Note how we request the name of the AvroTopic, and that we have to specify the serializer for the key and the serializer for the value for the Consumed class; in this case, the key is a String (always null), and the serializer for the value is our new avroSerde.

To deserealize the values for the HealthCheck Stream, we apply the same methods used in previous chapters inside the lambda of the mapValues() method with one argument (v->), shown as follows:

```
KStream healthCheckStream = avroStream.mapValues((v -> {
  GenericRecord healthCheckAvro = (GenericRecord) v;
  HealthCheck healthCheck = new HealthCheck(
    healthCheckAvro.get("event").toString(),
    healthCheckAvro.get("factory").toString(),
    healthCheckAvro.get("serialNumber").toString(),
    healthCheckAvro.get("type").toString(),
    healthCheckAvro.get("status").toString(),
    new Date((Long) healthCheckAvro.get("lastStartedAt")),
    Float.parseFloat(healthCheckAvro.get("temperature").toString()),
    healthCheckAvro.get("ipAddress").toString());
  return healthCheck;
}));
```

And again, the rest of the code of the process() method remains the same as in previous sections, shown as follows:

```
KStream uptimeStream = healthCheckStream.map(((KeyValueMapper)(k, v)-> {
  HealthCheck healthCheck = (HealthCheck) v;
  LocalDate startDateLocal = healthCheck.getLastStartedAt().toInstant()
              .atZone(ZoneId.systemDefault()).toLocalDate();
  int uptime =
    Period.between(startDateLocal, LocalDate.now()).getDays();
  return new KeyValue<>(
    healthCheck.getSerialNumber(), String.valueOf(uptime));
}));

uptimeStream.to( Constants.getUptimesTopic(),
    Produced.with(Serdes.String(), Serdes.String()));

Topology topology = streamsBuilder.build();
Properties props = new Properties();
```

```
props.put("bootstrap.servers", this.brokers);
props.put("application.id", "kioto");
KafkaStreams streams = new KafkaStreams(topology, props);
streams.start();
```

Note that the code could be cleaner: we could create our own Serde that includes the deserialization code, so we can directly deserialize Avro Objects into `HealthCheck` Objects. To achieve this, this class has to extend the generic Avro deserializer. We leave this as an exercise for you.

Running the AvroStreamsProcessor

To build the project, run this command from the `kioto` directory:

```
$ gradle build
```

If everything is correct, the output is something like the following:

```
BUILD SUCCESSFUL in 1s
  6 actionable task: 6 up-to-date
```

1. The first step is to run a console consumer for the `uptimes` topic, shown as follows:

   ```
   $ ./bin/kafka-console-consumer --bootstrap-server localhost:9092
   --topic uptimes --property print.key=true
   ```

2. From our IDE, run the main method of the `AvroStreamsProcessor`
3. From our IDE, run the main method of the `AvroProducer` (built in previous chapters)
4. The output on the console consumer for the `uptimes` topic should be similar to the following:

   ```
   EW05-HV36 33
   BO58-SB28 20
   DV03-ZT93 46
   . . .
   ```

Late event processing

Previously, we talked about message processing, but now we will talk about events. An event in this context is something that happens at a particular time. An event is a message that happens at a point in time.

In order to understand events, we have to know the timestamp semantics. An event always has two timestamps, shown as follows:

- **Event time**: The point in time when the event happened at the data source
- **Processing time**: The point in time when the event is processed in the data processor

Due to limitations imposed by the laws of physics, the processing time will always be subsequent to and necessarily different from the event time, for the following reasons:

- **There is always network latency**: The time to travel from the data source to the Kafka broker cannot be zero.
- **The client could have a cache**: If the client cached some events before, send them to the data processor. As an example, think about a mobile device that is not always connected to the network because there are zones without network access, and the device holds some data before sending it.
- **The existence of back pressure**: Sometimes, the broker will not process the events as they arrive, because it is busy and there are too many.

Having said the previous points, it is always important that our messages have a timestamp. Since version 0.10 of Kafka, the messages stored in Kafka always have an associated timestamp. The timestamp is normally assigned by the producer; if the producer sends a message without a timestamp, the broker assigns it one.

As a professional tip, when generating messages, always assign a timestamp from the producer.

Basic scenario

To explain late events, we need a system where the events arrive periodically and we want to know how many events are produced by unit of time. In *Figure 6.1*, we show this scenario:

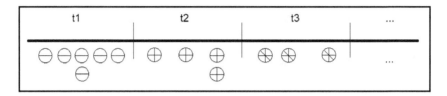

Figure 6.1: The events as they were produced

In the preceding figure, each marble represents an event. They are not supposed to have dimensions as they are at a specific point in time. Events are punctual, but for demonstration purposes, we represent them as balls. As we can see in **t1** and **t2,** two different events can happen at the same time.

In our figure, tn represents the nth time unit. Each marble represents a single event. To differentiate between them, the events on **t1** have one stripe, the events on **t2** have two stripes, and the events on **t3** have three stripes.

We want to count the events per unit of time, so we have the following:

- **t1** has six events
- **t2** has four events
- **t3** has three events

As systems have failures (such as network latency, shutdown of servers, network partitioning, power failures, voltage variations, and so on), suppose that an event that happened during **t2** has a delay and reached our system at **t3,** shown as follows in *Figure 6.2:*

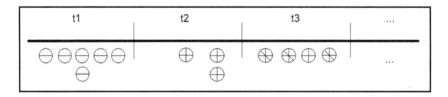

Figure 6.2: The events as they were processed

If we count our events using the **processing time**, we have the following results:

- **t1** has six events
- **t2** has three events
- **t3** has four events

If we have to calculate how many events were produced per time unit, our results would be incorrect.

The event that arrived on **t3** instead of **t2** is called a late event. We just have two alternatives, they are given as follows:

- When **t2** ends, produce a preliminary result that the count for **t2** is three events. And then, during processing, when we find in another time an event belonging to **t2**, we update the result for **t2**: **t2** has four events.
- When each window ends, we wait a little after the end before we produce a result. For example, we could wait another time unit. In this case, the results for tn are obtained when t(n+1) ends. Remember, the time to wait to produce results might not be related to the time unit size.

As you can guess, these scenarios are quite common in practice, and there are currently many interesting proposals. One of the most complete and advanced suites for handling late events is the Apache Beam proposal. However, Apache Spark, Apache Flink, and Akka Streams are also very powerful and attractive.

As we want to see how it is solved with Kafka Streams here, let's see that.

Late event generation

To test the Kafka Streams solution for late events, the first thing we need is a late event generator.

To simplify things, our generator will constantly send events at a fixed rate. And from time to time, it will generate a late event. The generator generates events with the following process:

- Each window is 10 seconds long
- It produces one event every second
- The event should be generated in 54th second of each minute, and will be delayed by 12 seconds; that is, it will arrive in the sixth second of the next minute (in the next window)

When we say that the window is of 10 seconds, we mean that we will make aggregations every 10 seconds. Remember that the objective of the test is that the late events are counted in the correct window.

Create the `src/main/java/kioto/events` directory and, inside it, create a file called `EventProducer.java` with the contents of *Listing 6.5,* shown as follows:

```
package kioto.events;
import ...
public final class EventProducer {
  private final Producer<String, String> producer;
  private EventProducer(String brokers) {
    Properties props = new Properties();
    props.put("bootstrap.servers", brokers);
    props.put("key.serializer", StringSerializer.class);
    props.put("value.serializer", StringSerializer.class);
    producer = new KafkaProducer<>(props);
  }
  private void produce() {
    // ...
  }
  private void sendMessage(long id, long ts, String info) {
    // ...
  }
  public static void main(String[] args) {
    (new EventProducer("localhost:9092")).produce();
  }
}
```

Listing 6.5: EventProducer.java

The event generator is a Java `KafkaProducer`, so declare the same properties as all the Kafka Producers.

The generator code is very simple, and the first thing that is required is a timer that generates an event every second. The timer triggers 0.3 seconds after every second to avoid messages sent at 0.998 seconds, for example. The `produce()` method is shown as follows:

```
private void produce() {
  long now = System.currentTimeMillis();
  long delay = 1300 - Math.floorMod(now, 1000);
  Timer timer = new Timer();
  timer.schedule(new TimerTask() {
    public void run() {
      long ts = System.currentTimeMillis();
      long second = Math.floorMod(ts / 1000, 60);
      if (second != 54) {
```

```
      EventProducer.this.sendMessage(second, ts, "on time");
    }
    if (second == 6) {
       EventProducer.this.sendMessage(54, ts - 12000, "late");
    }
  }
}, delay, 1000);
}
```

When the timer is triggered, the run method is executed. We send one event each second except on second 54, where we delay this event by 12 seconds. Then, we send this late event in the sixth second of the next minute, modifying the timestamp.

In the sendMessage() method, we just assign the timestamp of the event, shown as follows:

```
private void sendMessage(long id, long ts, String info) {
  long window = ts / 10000 * 10000;
  String value = "" + window + ',' + id + ',' + info;
  Future futureResult = this.producer.send(
     new ProducerRecord<>(
         "events", null, ts, String.valueOf(id), value));
  try {
    futureResult.get();
  } catch (InterruptedException | ExecutionException e) {
    // deal with the exception
  }
}
```

Running the EventProducer

To run the EventProducer, we follow these steps:

1. Create the events topic, as shown in the following block:

   ```
   $. /bin/kafka-topics --zookeeper localhost:2181 --create --topic
   events --replication-factor 1 --partitions 4
   ```

2. Run a console consumer for the events topic using the following command:

   ```
   $ ./bin/kafka-console-consumer --bootstrap-server localhost:9092
   --topic events
   ```

3. From the IDE, run the main method of the EventProducer.

4. The output on the console consumer for the events topic should be similar to the following:

```
1532529060000,47, on time
1532529060000,48, on time
1532529060000,49, on time
1532529070000,50, on time
1532529070000,51, on time
1532529070000,52, on time
1532529070000,53, on time
1532529070000,55, on time
1532529070000,56, on time
1532529070000,57, on time
1532529070000,58, on time
1532529070000,59, on time
1532529080000,0, on time
1532529080000,1, on time
1532529080000,2, on time
1532529080000,3, on time
1532529080000,4, on time
1532529080000,5, on time
1532529080000,6, on time
1532529070000,54, late
1532529080000,7, on time
. . .
```

Note that each event window changes every 10 seconds. Also, note how the 54th event is not sent between the 53rd and 55th events. The 54th event, belonging to a previous window, arrives in the next minute between the sixth and seventh seconds.

Kafka Streams processor

Now, let's solve the problem of counting how many events are in each window. For this, we will use Kafka Streams. When we do this type of analysis, it is said that we are doing **streaming aggregation**.

In the `src/main/java/kioto/events` directory, create a file called `EventProcessor.java` with the contents of *Listing 6.6*, shown as follows:

```
package kioto.events;
import ...
public final class EventProcessor {
  private final String brokers;
  private EventProcessor(String brokers) {
    this.brokers = brokers;
  }
  private void process() {
    // ...
  }
  public static void main(String[] args) {
    (new EventProcessor("localhost:9092")).process();
  }
}
```

Listing 6.6: EventProcessor.java

All the processing logic is contained in the process() method. The first step is to create a StreamsBuilder to create the KStream, shown as follows:

```
StreamsBuilder streamsBuilder = new StreamsBuilder();
KStream stream = streamsBuilder.stream(
  "events", Consumed.with(Serdes.String(), Serdes.String()));
```

As we know, we specify from topic we are reading the events in this case is called **events**, and then we always specify the Serdes, both keys and values of type String.

If you remember, we have each step as a transformation from one stream to another.

The next step is to build a KTable. To do so, we first use the groupBy() function, which receives a key-value pair, and we assign a key called "foo", because it is not relevant but we need to specify one. Then, we apply the windowedBy() function, specifying that the window will be 10 seconds long. Finally, we use the count() function, so we are producing key-value pairs with String as keys and long as values. This number is the count of the events for each window (the key is the window start time):

```
KTable aggregates = stream
  .groupBy( (k, v) -> "foo", Serialized.with(Serdes.String(),
Serdes.String()))
  .windowedBy( TimeWindows.of(10000L) )
  .count( Materialized.with( Serdes.String(), Serdes.Long() ) );
```

If you have problems with the conceptual visualization of the `KTable`, which keys are of type `KTable<Windowed<String>>` and values are of type `long`, and printing it (in the KSQL chapter we will see how to do it), would be something like the one, as follows:

```
key | value
----------------- |-------
1532529050000:foo | 10
1532529060000:foo | 10
1532529070000:foo | 9
1532529080000:foo | 3
. . .
```

The key has the window ID and the utility aggregation key with value `"foo"`. The value is the number of elements counted in the window at a specific point of time.

Next, as we need to output the `KTable` to a topic, we need to convert it to a `KStream` as follows:

```
aggregates
  .toStream()
  .map( (ws, i) -> new KeyValue( ""+((Windowed)ws).window().start(), ""+i))
  .to("aggregates", Produced.with(Serdes.String(), Serdes.String())));
```

The `toStream()` method of the `KTable` returns a `KStream`. We use a `map()` function that receives two values, the window and the count, then we extract the window start time as the key and the count as the value. The `to()` method specifies to which topic we want to output (always specifying the serdes as a good practice).

Finally, as in previous sections, we need to start the topology and the application, shown as follows:

```
Topology topology = streamsBuilder.build();
Properties props = new Properties();
props.put("bootstrap.servers", this.brokers);
props.put("application.id", "kioto");
props.put("auto.offset.reset", "latest");
props.put("commit.interval.ms", 30000);
KafkaStreams streams = new KafkaStreams(topology, props);
streams.start();
```

Remember that the `commit.interval.ms` property indicates how many milliseconds we will wait to write the results to the `aggregates` topic.

Running the Streams processor

To run the `EventProcessor`, follow these steps:

1. Create the `aggregates` topic as follows:

   ```
   $. /bin/kafka-topics --zookeeper localhost:2181 --create --topic
   aggregates --replication-factor 1 --partitions 4
   ```

2. Run a console consumer for the aggregates topic, as follows:

   ```
   $ ./bin/kafka-console-consumer --bootstrap-server localhost:9092
   --topic aggregates --property print.key=true
   ```

3. From the IDE, run the main method of the `EventProducer`.
4. From the IDE, run the main method of the `EventProcessor`.
5. Remember that it writes to the topic every 30 seconds. The output on the console consumer for the aggregates topic should be similar to the following:

   ```
   1532529050000  10
   1532529060000  10
   1532529070000  9
   1532529080000  3
   ```

   After the second window, we can see that the values in the `KTable` are updated with fresh (and correct) data, shown as follows:

   ```
   1532529050000  10
   1532529060000  10
   1532529070000  10
   1532529080000  10
   1532529090000  10
   1532529100000  4
   ```

Note how in the first print, the value for the last window is 3, and the window started in `1532529070000` has a value of `9`. Then in the second print, the values are correct. This behavior is because in the first print, the delayed event had not arrived yet. When this finally arrived, the count values were corrected for all the windows.

Stream processor analysis

If you have a lot of questions here, it is normal.

The first thought to consider is that in streaming aggregation, and in streaming in general, the Streams are unbounded. It is never clear when we will take the final results, that is, we as programmers have to decide when to consider a partial value of an aggregation as a final result.

Recall that the print of the Stream is an instant photo of the KTable at a certain time. Therefore, the results of a KTable are only valid at the time of the output. It is important to remember that in the future, the values of the KTable may be different. Now, to see results more frequently, change the value of the commit interval to zero, shown as follows:

```
props.put("commit.interval.ms", 0);
```

This line says that the results of the KTable will be printed when they are modified, that is, it will print new values every second. If you run the program, the value of the KTable will be printed with each update (every second), shown as follows:

```
1532529080000 6
1532529080000 7
1532529080000 8
1532529080000 9
1532529080000 10 <-- Window end
1532529090000 1  <-- Window beginning
1532529090000 2
1532529090000 3
1532529090000 5  <-- The 4th didn't arrive
1532529090000 6
1532529090000 7
1532529090000 8
1532529090000 9  <-- Window end
1532529100000 1
1532529100000 2
1532529100000 3
1532529100000 4
1532529100000 5
1532529100000 6
1532529090000 10 <-- The 4th arrived, so the count value is updated
1532529100000 7
1532529100000 8
. . .
```

Keep a note of two effects:

- The aggregate result (the count) for the window stops at 9 when the window ends and the next window events begin to arrive
- When the late event finally arrives, it produces an update in the window's count

Yes, Kafka Streams apply event time semantics in order to do the aggregation. It is important to remember that in order to visualize the data, we had to modify the commit interval. Leaving this value at zero would have negative repercussions on a production environment.

As you may guess, processing an event stream is much more complex than processing a fixed dataset. The events usually arrive late, in disorder, and it is practically impossible to know when the totality of the data has arrived. How do you know when there are late events? If there is, how much should we expect for them? When should we discard a late event?

The quality of a programmer is determined by the quality of their tools. The capabilities of the processing tool make a big difference when processing data. In this context, we have to reflect when the results are produced and when they arrived late.

The process of discarding events has a special name: watermarking. In Kafka Streams, this is achieved through setting the aggregation windows' retention period.

Summary

Kafka Streams is a powerful library, and is the only option when building data pipelines with Apache Kafka. Kafka Streams removes much of the boilerplate work needed when implementing plain Java clients. Compared to Apache Spark or Apache Flink, the Kafka Streams applications are much simpler to build and manage.

We also have seen how to improve a Kafka Streams application to deserialize data in JSON and Avro formats. The serialization part (writing to a topic) is very similar since we are using SerDes that are capable of both data serialization and deserialization.

For those who work with Scala, there is a library for Kafka Streams called circe that offers SerDes to manipulate JSON data. The circe library is the equivalent in Scala of the Jackson library.

As mentioned earlier, Apache Beam has a more complex suite of tools, but is totally focused on Stream management. Its model is based on triggers and semantics between events. It also has a powerful model for watermark handling.

One notable advantage of Kafka Streams over Apache Beam is that its deployment model is simpler. This leads many developers to adopt it. However, for more complex problems, Apache Beam may be a better tool.

In the following chapters, we will talk about how to get the best of two worlds: Apache Spark and Kafka Streams.

7
KSQL

In previous chapters, we wrote Java code to manipulate data streams with Kafka, and we also we built several Java processors for Kafka and Kafka Streams. In this chapter, we will use KSQL to achieve the same results.

This chapter covers the following topics:

- KSQL in a nutshell
- Running KSQL
- Using the KSQL CLI
- Processing data with KSQL
- Writing to a topic

KSQL in a nutshell

With Kafka Connect, we can build clients in several programming languages: JVM (Java, Clojure, Scala), C/C++, C#, Python, Go, Erlang, Ruby, Node.js, Perl, PHP, Rust, and Swift. In addition to this, if your programming language is not listed, you can use the Kafka REST proxy. But the Kafka authors realized that all programmers, especially data engineers, can all talk the same language: **Structured Query Language** (**SQL**). So, they decided to create an abstraction layer on Kafka Streams in which they could manipulate and query streams using SQL.

KSQL is a SQL engine for Apache Kafka. It allows writing SQL sentences to analyze data streams in real time. Remember that a stream is an unbounded data structure, so we don't know where it begins, and we are constantly receiving new data. Therefore, KSQL queries usually keep generating results until you stop them.

KSQL runs over Kafka Streams. To run queries over a data stream, the queries are parsed, analyzed, and then a Kafka Streams topology is built and executed, just as we did at the end of each `process()` method when running Kafka Streams applications. KSQL has mapped the Kafka Streams concepts one to one, for example, tables, joins, streams, windowing functions, and so on.

KSQL runs on KSQL servers. So if we need more capacity, we can run one or more instances of KSQL servers. Internally, all the KSQL instances work together, sending and receiving information through a dedicated and private topic called `_confluent-ksql-default__command_topic`.

As with all Kafka technologies, we can also interact with KSQL through a REST API. Also, KSQL, has its own fancy **command-line interface** (**CLI**). If you want to read more about KSQL read the online documentation at the following URL: `https://docs.confluent.io/current/ksql/docs/index.html`.

Running KSQL

As mentioned previously, KSQL is shipped with the Confluent Platform. When we start the Confluent Platform, automatically at the end it starts a KSQL server, as shown in the *Figure 7 .1*:

```
./confluent start

This CLI is intended for development only, not for production
https://docs.confluent.io/current/cli/index.html

Starting zookeeper
zookeeper is [UP]
Starting kafka
kafka is [UP]
Starting schema-registry
schema-registry is [UP]
Starting kafka-rest
kafka-rest is [UP]
Starting connect
connect is [UP]
Starting ksql-server
ksql-server is [UP]
```

Figure 7.1: Confluent Platform startup

To start the KSQL server alone (not recommendable), we can use the `ksql-server-start` command. Just type `./ksql` from the bin directory, as shown in *Figure 7.2*:

Figure 7.2: KSQL CLI start screen

Using the KSQL CLI

The KSQL CLI is a command prompt to interact with KSQL; it is very similar to the one that comes with relational databases such as MariaDB or MySQL. To see all the possible commands, type help and a list with the options will be displayed.

At the moment, we have not informed KSQL of anything. We must declare that something is a table or a stream. We will use the information produced from previous chapters with the producers that write JSON information to the `healthchecks` topic.

If you remember, the data looks like this:

```
{"event":"HEALTH_CHECK","factory":"Lake Anyaport","serialNumber":"EW05-
HV36","type":"WIND","status":"STARTING","lastStartedAt":"2018-09-17T11:05:2
6.094+0000","temperature":62.0,"ipAddress":"15.185.195.90"}
{"event":"HEALTH_CHECK","factory":"Candelariohaven","serialNumber":"BO58-
SB28","type":"SOLAR","status":"STARTING","lastStartedAt":"2018-08-16T04:00:
```

```
00.179+0000","temperature":75.0,"ipAddress":"151.157.164.162"}
{"event":"HEALTH_CHECK","factory":"Ramonaview","serialNumber":"DV03-
ZT93","type":"SOLAR","status":"RUNNING","lastStartedAt":"2018-07-12T10:16:3
9.091+0000","temperature":70.0,"ipAddress":"173.141.90.85"}
. . .
```

KSQL can read JSON data, and can also read data in Avro format. To declare a stream from the `healthchecks` topic, we use the following command:

```
ksql>  CREATE STREAM healthchecks (event string, factory string,
serialNumber string, type string, status string, lastStartedAt string,
temperature double, ipAddress string) WITH (kafka_topic='healthchecks',
value_format='json');
```

The output is similar to this:

```
Message
---------------------------
Stream created and running
---------------------------
```

To review the structure of an existing `STREAM`, we can use the `DESCRIBE` command, which is shown here and that tells us the data types and their structure:

```
ksql> DESCRIBE healthchecks;
```

The output is similar to the following:

```
Name          : HEALTHCHECKS
Field         | Type
-------------------------------------------------
ROWTIME       | BIGINT            (system)
ROWKEY        | VARCHAR(STRING)   (system)
EVENT         | VARCHAR(STRING)
FACTORY       | VARCHAR(STRING)
SERIALNUMBER  | VARCHAR(STRING)
TYPE          | VARCHAR(STRING)
STATUS        | VARCHAR(STRING)
LASTSTARTEDAT | VARCHAR(STRING)
TEMPERATURE   | DOUBLE
IPADDRESS     | VARCHAR(STRING)
```

Note that at the beginning, two extra fields are shown: `ROWTIME` (the message timestamp) and `ROWKEY` (the message key).

When we created the stream, we declared that the Kafka topic is `healthchecks`. So, if we execute the `SELECT` command, we obtain a list of the events that are in the topic to which our stream points in real time (remember to run a producer to obtain fresh data). The command is as follows:

```
ksql> select * from healthchecks;
```

The output is similar to this:

```
1532598615943 | null | HEALTH_CHECK | Carliefort | FM41-RE80 | WIND |
STARTING | 2017-08-13T09:37:21.681+0000 | 46.0 | 228.247.233.14
1532598616454 | null | HEALTH_CHECK | East Waldo | HN72-EB29 | WIND |
RUNNING | 2017-10-31T14:20:13.929+0000 | 3.0 | 223.5.127.146
1532598616961 | null | HEALTH_CHECK | New Cooper | MM04-TZ21 | SOLAR |
SHUTTING_DOWN | 2017-08-21T21:10:31.190+0000 | 23.0 | 233.143.140.46
1532598617463 | null | HEALTH_CHECK | Mannmouth | XM02-PQ43 | GEOTHERMAL |
RUNNING | 2017-09-08T10:44:56.005+0000 | 73.0 | 221.96.17.237
1532598617968 | null | HEALTH_CHECK | Elvisfort | WP70-RY81 | NUCLEAR |
RUNNING | 2017-09-07T02:40:18.917+0000 | 49.0 | 182.94.17.58
1532598618475 | null | HEALTH_CHECK | Larkinstad | XD75-FY56 | GEOTHERMAL |
STARTING | 2017-09-06T08:48:14.139+0000 | 35.0 | 105.236.9.137
1532598618979 | null | HEALTH_CHECK | Nakiaton | BA85-FY32 | SOLAR |
RUNNING | 2017-08-15T04:10:02.590+0000 | 32.0 | 185.210.26.215
1532598619483 | null | HEALTH_CHECK | North Brady | NO31-LM78 |
HYDROELECTRIC | RUNNING | 2017-10-05T12:12:52.940+0000 | 5.0 | 17.48.190.21
1532598619989 | null | HEALTH_CHECK | North Josianemouth | GT17-TZ11 |
SOLAR | SHUTTING_DOWN | 2017-08-29T16:57:23.000+0000 | 6.0 | 99.202.136.163
```

The `SELECT` command shows the data from the Kafka topic declared in the stream. The query never stops, so it will run till you stop it. New records are printed as new lines, as new events are produced in the topic. To stop a query, type *Ctrl + C*.

Processing data with KSQL

In previous chapters, we took the data from the `healthchecks` topic, calculated the `uptimes` of the machines, and pushed this data into a topic called `uptimes`. Now, we are going to do this with KSQL.

At the time of writing, KSQL does not yet have a function to compare two dates, so we have the following two options:

- Code a **user-defined function** (UDF) for KSQL in Java
- Use the existing functions to make our calculation

As creating a new UDF is out of scope for now, let's go for the second option: use the existing functions to make our calculation.

The first step is to parse the startup time using the STRINGTOTIMESTAMP function, shown as follows (remember that we declared the date in string format, because KSQL doesn't yet have a DATE type):

```
ksql> SELECT event, factory, serialNumber, type, status, lastStartedAt,
temperature, ipAddress, STRINGTOTIMESTAMP(lastStartedAt, 'yyyy-MM-
dd''T''HH:mm:ss.SSSZ') FROM healthchecks;
```

The output is similar to this:

```
HEALTH_CHECK | Ezekielfurt | AW90-DQ16 | HYDROELECTRIC | RUNNING |
2017-09-28T21:00:45.683+0000 | 7.0 | 89.87.184.250 | 1532168445683
HEALTH_CHECK | Icieville | WB52-WC16 | WIND | SHUTTING_DOWN |
2017-10-31T22:38:26.783+0000 | 15.0 | 40.23.168.167 | 1532025506783
HEALTH_CHECK | McClurehaven | QP68-WX17 | GEOTHERMAL | RUNNING |
2017-11-12T23:16:27.105+0000 | 76.0 | 252.213.150.75 | 1532064587105
HEALTH_CHECK | East Maudshire | DO15-BB56 | NUCLEAR | STARTING |
2017-10-14T03:04:00.399+0000 | 51.0 | 93.202.28.134 | 1532486240399
HEALTH_CHECK | South Johnhaven | EE06-EX06 | HYDROELECTRIC | RUNNING |
2017-09-06T20:14:27.438+0000 | 91.0 | 244.254.181.218 | 1532264867438
```

The next step is to compare these dates to the current date. In KSQL, at the moment, there is no function to get today's date either, so let's use the STRINGTOTIMESTAMP function to parse today's date, shown as follows:

```
ksql> SELECT serialNumber, STRINGTOTIMESTAMP(lastStartedAt, 'yyyy-MM-
dd''T''HH:mm:ss.SSSZ'), STRINGTOTIMESTAMP('2017-11-18', 'yyyy-MM-dd') FROM
healthchecks;
```

The output is similar to this:

```
FE79-DN10 | 1532050647607 | 1510984800000
XE79-WP47 | 1532971000830 | 1510984800000
MP03-XC09 | 1532260107928 | 1510984800000
SO48-QF28 | 1532223768121 | 1510984800000
OC25-AB61 | 1532541923073 | 1510984800000
AL60-XM70 | 1532932441768 | 1510984800000
```

Now, let's compare these two dates and calculate the number of days between them, shown as follows (1 day = 86,400 seconds = 24 hours x 60 minutes x 60 seconds, 1 second = 1,000 milliseconds):

```
ksql> SELECT serialNumber, (STRINGTOTIMESTAMP('2017-11-18','yyyy-MM-
dd''T''HH:mm:ss.SSSZ')-STRINGTOTIMESTAMP(lastStartedAt,'yyyy-MM-
dd'))/86400/1000 FROM healthchecks;
```

The output is similar to this:

```
EH92-AQ09 | 39
BB09-XG98 | 42
LE94-BT50 | 21
GO25-IE91 | 97
WD93-HP20 | 22
JX48-KN03 | 12
EC84-DD11 | 73
SF06-UB22 | 47
IU77-VQ89 | 18
NM80-ZY31 | 5
TR64-TI21 | 51
ZQ13-GI11 | 80
II04-MB66 | 48
```

Perfect, now we have calculated the uptime for every machine.

Writing to a topic

So far, we have processed the data and printed the results in real time. To send these results to another topic, we use a CREATE command modality, where it is specified from a SELECT.

Let's start by writing the uptime as a string and writing the data in a comma-delimited format, shown as follows (remember that KSQL supports comma-delimited, JSON, and Avro formats). At the moment, it's enough because we're only writing one value:

```
ksql> CREATE STREAM uptimes WITH (kafka_topic='uptimes',
value_format='delimited') AS SELECT
CAST((STRINGTOTIMESTAMP('2017-11-18','yyyy-MM-dd''T''HH:mm:ss.SSSZ')-
STRINGTOTIMESTAMP(lastStartedAt,'yyyy-MM-dd'))/86400/1000 AS string) AS
uptime FROM healthchecks;
```

The output is similar to this:

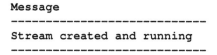

```
Message
---------------------------------
Stream created and running
---------------------------------
```

Our query is running in the background. To see it is running, we could use a console consumer of the `uptimes` topic, shown as follows:

```
$ ./kafka-console-consumer --bootstrap-server localhost:9092 --topic
uptimes --property print.key=true
```

The output is similar to this:

```
null   39
null   42
null   21
```

The results are correct; however, we forgot to use the machine serial number as the message key. To do this, we have to rebuild our query and our stream.

The first step is to use the `show queries` command, shown here:

```
ksql> show queries;
```

The output is similar to this:

```
 Query ID        | Kafka Topic | Query String
-----------------------------------------------------------------------
----
CSAS_UPTIMES_0 | UPTIMES     | CREATE STREAM uptimes WITH
(kafka_topic='uptimes', value_format='delimited') AS SELECT
CAST((STRINGTOTIMESTAMP('2017-11-18','yyyy-MM-dd''T''HH:mm:ss.SSSZ')-
STRINGTOTIMESTAMP(lastStartedAt,'yyyy-MM-dd'))/86400/1000 AS string) AS
uptime FROM healthchecks;
-----------------------------------------------------------------------
----
For detailed information on a Query run: EXPLAIN <Query ID>;
```

With the `Query ID`, use the `terminate <ID>` command, shown as follows:

```
ksql> terminate CSAS_UPTIMES_0;
```

The output is similar to this:

```
Message
------------------
Query terminated.
------------------
```

To delete the stream, use the `DROP STREAM` command, shown as follows:

```
ksql> DROP STREAM uptimes;
```

The output is similar to this:

```
Message
-------------------------------
Source UPTIMES was dropped.
-------------------------------
```

To write the events key correctly, we must use the PARTITION BY clause. First, we regenerate our stream with a partial calculation, shown as follows:

```
ksql> CREATE STREAM healthchecks_processed AS SELECT serialNumber,
CAST((STRINGTOTIMESTAMP('2017-11-18','yyyy-MM-dd''T''HH:mm:ss.SSSZ')-
STRINGTOTIMESTAMP(lastStartedAt,'yyyy-MM-dd'))/86400/1000 AS string) AS
uptime FROM healthchecks;
```

The output is similar to this:

```
Message
-------------------------------
Stream created and running
-------------------------------
```

This stream has two fields (serialNumber and uptime). To write these calculated values to a topic, we use CREATE STREAM, AS SELECT as follows:

```
ksql> CREATE STREAM uptimes WITH (kafka_topic='uptimes',
value_format='delimited') AS SELECT * FROM healthchecks_processed;
```

The output is similar to this:

```
Message
-------------------------------
Stream created and running
-------------------------------
```

Finally, run a console consumer to show the results, demonstrated as follows:

```
$ ./bin/kafka-console-consumer --bootstrap-server localhost:9092 --topic
uptimes --property print.key=true
```

The output is similar to this:

```
EW05-HV36    33
BO58-SB28    20
DV03-ZT93    46
...
```

Now, close the KSQL CLI (*Ctrl* + *C* and close the command window). As the queries are still running in KSQL, you still see the outputs in the console consumer window.

Congratulations, you have built a Kafka Streams application with a few KSQL commands.

To unveil all the power of KSQL, it is important to review the official documentation at the following address:

```
https://docs.confluent.io/current/ksql/docs/tutorials/index.html
```

Summary

KSQL is still very new, but the product has gained adoption among developers. We all hope it continues to be extended to support more data formats (Protobuffers, Thrift, and so on) and more functions (more UDFs, such as geolocation and IoT, that are quite useful).

So again, congratulations! In this chapter, we did the same as in the previous ones, but without writing a single line of Java code. This makes KSQL the preferred tool for people who are not programmers, but are dedicated to data analysis.

8
Kafka Connect

In this chapter, instead of using the Kafka Java API for producers and consumers, Kafka Streams, or KSQL as in previous chapters, we are going to connect Kafka with Spark Structured Streaming, the Apache Spark solution to process streams with its Datasets API.

This chapter covers the following topics:

- Spark Streaming processor
- Reading Kafka from Spark
- Data conversion
- Data processing
- Writing to Kafka from Spark
- Running the `SparkProcessor`

Kafka Connect in a nutshell

Kafka Connect is an open source framework, part of Apache Kafka; it is used to connect Kafka with other systems, such as structured databases, column stores, key-value stores, filesystems, and search engines.

Kafka Connect has a wide range of built-in connectors. If we are reading from the external system, it is called a **data source**; if we are writing to the external system, it is called a **data sink**.

In previous chapters, we created a Java Kafka Producer that sends JSON data to a topic called `healthchecks` in messages like these three:

```
{"event":"HEALTH_CHECK","factory":"Lake Anyaport","serialNumber":"EW05-
HV36","type":"WIND","status":"STARTING","lastStartedAt":"2018-09-17T11:05:2
6.094+0000","temperature":62.0,"ipAddress":"15.185.195.90"}
{"event":"HEALTH_CHECK","factory":"Candelariohaven","serialNumber":"BO58-
SB28","type":"SOLAR","status":"STARTING","lastStartedAt":"2018-08-16T04:00:
```

```
00.179+0000","temperature":75.0,"ipAddress":"151.157.164.162"}{"event":"HEA
LTH_CHECK","factory":"Ramonaview","serialNumber":"DV03-
ZT93","type":"SOLAR","status":"RUNNING","lastStartedAt":"2018-07-12T10:16:3
9.091+0000","temperature":70.0,"ipAddress":"173.141.90.85"}
. . .
```

Now, we are going to process this data to calculate the machine's uptime and to obtain a topic with messages like these three:

```
EW05-HV36    33
BO58-SB28    20
DV03-ZT93    46
. . .
```

Project setup

The first step is to modify our Kioto project. We have to add the dependencies to `build.gradle`, as shown in *Listing 8.1*:

```
apply plugin: 'java'
apply plugin: 'application'
sourceCompatibility = '1.8'
mainClassName = 'kioto.ProcessingEngine'
repositories {
    mavenCentral()
    maven { url 'https://packages.confluent.io/maven/' }
}
version = '0.1.0'
dependencies {
    compile 'com.github.javafaker:javafaker:0.15'
    compile 'com.fasterxml.jackson.core:jackson-core:2.9.7'
    compile 'io.confluent:kafka-avro-serializer:5.0.0'
    compile 'org.apache.kafka:kafka_2.12:2.0.0'
    compile 'org.apache.kafka:kafka-streams:2.0.0'
    compile 'io.confluent:kafka-streams-avro-serde:5.0.0'
    compile 'org.apache.spark:spark-sql_2.11:2.2.2'
    compile 'org.apache.spark:spark-sql-kafka-0-10_2.11:2.2.2'
}
jar {
    manifest {
        attributes 'Main-Class': mainClassName
    } from {
        configurations.compile.collect {
            it.isDirectory() ? it : zipTree(it)
        }
    }
```

```
    exclude "META-INF/*.SF"
    exclude "META-INF/*.DSA"
    exclude "META-INF/*.RSA"
}
```

Listing 8.1: Kioto gradle build file for Spark

To use Apache Spark, we need the dependency, shown as follows:

```
compile 'org.apache.spark:spark-sql_2.11:2.2.2'
```

To connect Apache Spark with Kafka, we need the dependency, shown as follows:

```
compile 'org.apache.spark:spark-sql-kafka-0-10_2.11:2.2.2'
```

We are using an old Spark version, 2.2.2, for the following two reasons:

- At the moment you are reading this, surely the Spark version will be superior. The reason why I chose this version (and not the last one at the time of writing) is because the connector with Kafka works perfectly with this version (in performance and with regard to bugs).
- The Kafka connector that works with this version is several versions behind the most modern version of the Kafka connector. You always have to consider this when upgrading production environments.

Spark Streaming processor

Now, in the `src/main/java/kioto/spark` directory, create a file called `SparkProcessor.java` with the contents of *Listing 8.2*, shown as follows:

```java
package kioto.spark;
import kioto.Constants;
import org.apache.spark.sql.*;
import org.apache.spark.sql.streaming.*;
import org.apache.spark.sql.types.*;
import java.sql.Timestamp;
import java.time.LocalDate;
import java.time.Period;

public class SparkProcessor {
  private String brokers;
  public SparkProcessor(String brokers) {
    this.brokers = brokers;
  }
  public final void process() {
```

```
    //below is the content of this method
  }
  public static void main(String[] args) {
    (new SparkProcessor("localhost:9092")).process();
  }
}
```

Listing 8.2: SparkProcessor.java

Note that, as in previous examples, the main method invoked the `process()` method with the IP address and the port of the Kafka brokers.

Now, let's fill the `process()` method. The first step is to initialize Spark, as demonstrated in the following block:

```
SparkSession spark = SparkSession.builder()
    .appName("kioto")
    .master("local[*]")
    .getOrCreate();
```

In Spark, the application name must be the same for each member in the cluster, so here we call it Kioto (original, isn't it?).

As we are going to run the application locally, we are setting the Spark master to `local[*]`, which means that we are creating a number of threads equivalent to the machine CPU cores.

Reading Kafka from Spark

There are several connectors for Apache Spark. In this case, we are using the Databricks Inc. (the company responsible for Apache Spark) connector for Kafka.

Using this Spark Kafka connector, we can read data with Spark Structured Streaming from a Kafka topic:

```
Dataset<Row> inputDataset = spark
    .readStream()
    .format("kafka")
    .option("kafka.bootstrap.servers", brokers)
    .option("subscribe", Constants.getHealthChecksTopic())
    .load();
```

Simply by saying Kafka format, we can read a stream from the topic specified in the `subscribe` option, running on the brokers specified.

At this point in the code, if you invoke the `printSchema()` method on the `inputDataSet`, the result will be something similar to *Figure 8.1*:

```
root
 |-- key: binary (nullable = true)
 |-- value: binary (nullable = true)
 |-- topic: string (nullable = true)
 |-- partition: integer (nullable = true)
 |-- offset: long (nullable = true)
 |-- timestamp: timestamp (nullable = true)
 |-- timestampType: integer (nullable = true)
```

Figure 8.1: Print schema output

We can interpret this as follows:

- The key and the value are binary data. Here in Spark, unfortunately, and unlike Kafka, it is not possible to specify deserializers for our data. So, it is necessary to use Dataframe operations to do the deserialization.
- For each message, we can know the topic, the partition, the offset, and the timestamp.
- The timestamp type is always zero.

As with Kafka Streams, with Spark Streaming, in each step we have to generate a new data stream in order to apply transformations and get new ones.

In each step, if we need to print our data stream (to debug the application), we can use the following code:

```
StreamingQuery consoleOutput =
    streamToPrint.writeStream()
    .outputMode("append")
    .format("console")
    .start();
```

The first line is optional, because we really don't need to assign the result to an object, just the code execution.

The output of this snippet is something like *Figure 8.2*. The message value is certainly binary data:

```
-------------------------------------------
Batch: 3
-------------------------------------------
+----+--------------------+------------+---------+------+--------------------+-------------+
| key|               value|       topic|partition|offset|           timestamp|timestampType|
+----+--------------------+------------+---------+------+--------------------+-------------+
|null|[32 7C 55 29 45 6...|healthchecks|        0| 25462|2017-07-11 09:45:...|            0|
+----+--------------------+------------+---------+------+--------------------+-------------+

-------------------------------------------
Batch: 4
-------------------------------------------
+----+--------------------+------------+---------+------+--------------------+-------------+
| key|               value|       topic|partition|offset|           timestamp|timestampType|
+----+--------------------+------------+---------+------+--------------------+-------------+
|null|[32 7C 55 29 45 6...|healthchecks|        0| 25463|2017-07-11 09:45:...|            0|
+----+--------------------+------------+---------+------+--------------------+-------------+
```

Figure 8.2: Data stream console output

Data conversion

We know that when we produced the data, it was in JSON format, although Spark reads it in binary format. To convert the binary message to string, we use the following code:

```
Dataset<Row> healthCheckJsonDf =
    inputDataset.selectExpr("CAST(value AS STRING)");
```

The `Dataset` console output is now human-readable, and is shown as follows:

```
+--------------------------+
|                     value|
+--------------------------+
| {"event":"HEALTH_CHECK...|
+--------------------------+
```

The next step is to provide the fields list to specify the data structure of the JSON message, as follows:

```
StructType struct = new StructType()
    .add("event", DataTypes.StringType)
    .add("factory", DataTypes.StringType)
    .add("serialNumber", DataTypes.StringType)
    .add("type", DataTypes.StringType)
    .add("status", DataTypes.StringType)
    .add("lastStartedAt", DataTypes.StringType)
    .add("temperature", DataTypes.FloatType)
    .add("ipAddress", DataTypes.StringType);
```

Next, we deserialize the String in JSON format. The simplest way is to use the prebuilt `from_json()` function in the `org.apache.spark.sql.functions` package, which is demonstrated in the following block:

```
Dataset<Row> healthCheckNestedDs =
    healthCheckJsonDf.select(
        functions.from_json(
            new Column("value"), struct).as("healthCheck"));
```

If we print the `Dataset` at this point, we can see the columns nested as we indicated in the schema:

```
root
 |-- healthcheck: struct (nullable = true)
 |    |-- event: string (nullable = true)
 |    |-- factory: string (nullable = true)
 |    |-- serialNumber: string (nullable = true)
 |    |-- type: string (nullable = true)
 |    |-- status: string (nullable = true)
 |    |-- lastStartedAt: string (nullable = true)
 |    |-- temperature: float (nullable = true)
 |    |-- ipAddress: string (nullable = true)
```

The next step is to flatten this `Dataset`, as follows:

```
Dataset<Row> healthCheckFlattenedDs = healthCheckNestedDs
    .selectExpr("healthCheck.serialNumber", "healthCheck.lastStartedAt");
```

To visualize the flattening, if we print the `Dataset`, we get the following:

```
root
 |-- serialNumber: string (nullable = true)
 |-- lastStartedAt: string (nullable = true)
```

Note that we read the startup time as a string. This is because internally the `from_json()` function uses the Jackson library. Unfortunately, there is no way to specify the format of the date to be read.

For these purposes, fortunately there is the `to_timestamp()` function in the same functions package. There is also the `to_date()` function if it is necessary to read only a date, ignoring the time specification. Here, we are rewriting the `lastStartedAt` column, similar to this:

```
Dataset<Row> healthCheckDs = healthCheckFlattenedDs
    .withColumn("lastStartedAt", functions.to_timestamp(
        new Column ("lastStartedAt"), "yyyy-MM-dd'T'HH:mm:ss.SSSZ"));
```

Data processing

Now, what we are going to do is to calculate the `uptimes`. As is to be expected, Spark does not have a built-in function to calculate the number of days between two dates, so we are going to create a user-defined function.

If we remember the KSQL chapter, it is also possible to build and use new UDFs in KSQL.

To achieve this, the first thing we do is build a function that receives as input a `java.sql.Timestamp`, as shown in the following code (this is how timestamps are represented in the Spark DataSets) and returns an integer with the number of days from that date:

```
private final int uptimeFunc(Timestamp date) {
    LocalDate localDate = date.toLocalDateTime().toLocalDate();
    return Period.between(localDate, LocalDate.now()).getDays();
}
```

The next step is to generate a Spark UDF as follows:

```
Dataset<Row> processedDs = healthCheckDs
    .withColumn( "lastStartedAt", new Column("uptime"));
```

And finally, apply that UDF to the `lastStartedAt` column to create a new column in the `Dataset` called `uptime`.

Writing to Kafka from Spark

As we already processed the data and calculated the `uptime`, now all we need to do is to write these values in the Kafka topic called `uptimes`.

Kafka's connector allows us to write values to Kafka. The requirement is that the `Dataset` to write must have a column called `key` and another column called `value`; each one can be of the type String or binary.

Since we want the machine serial number to be the key, there is no problem if it is already of String type. Now, we just have to convert the `uptime` column from binary into String.

We use the `select()` method of the `Dataset` class to calculate these two columns and assign them new names using the `as()` method, shown as follows (to do this, we could also use the `alias()` method of that class):

```
Dataset<Row> resDf = processedDs.select(
    (new Column("serialNumber")).as("key"),
    processedDs.col("uptime").cast(DataTypes.StringType).as("value"));
```

Our `Dataset` is ready and it has the format expected by the Kafka connector. The following code is to tell Spark to write these values to Kafka:

```
//StreamingQuery kafkaOutput =
resDf.writeStream()
    .format("kafka")
    .option("kafka.bootstrap.servers", brokers)
    .option("topic", "uptimes")
    .option("checkpointLocation", "/temp")
    .start();
```

Note that we added the checkpoint location in the options. This is to ensure the high availability of Kafka. However, this does not guarantee that messages are delivered in exactly once mode. Nowadays, Kafka can guarantee exactly once delivery; Spark for the moment, can only guarantee the at least once delivery mode.

Finally, we call the `awaitAnyTermination()` method, shown as follows:

```
try {
  spark.streams().awaitAnyTermination();
} catch (StreamingQueryException e) {
  // deal with the Exception
}
```

An important note is to mention that if Spark leaves a console output inside the code, it implies that all queries must call its `start()` method before calling any `awaitTermination()` method, shown as follows:

```
firstOutput = someDataSet.writeStream
...
    .start()
...
 secondOutput = anotherDataSet.writeStream
...
    .start()
firstOutput.awaitTermination()
anotherOutput.awaitTermination()
```

Also note that we can replace all the `awaitTermination()` calls at the end with a single call to `awaitAnyTermination()`, as we did in the original code.

Running the SparkProcessor

To build the project, run this command from the `kioto` directory as follows:

```
$ gradle jar
```

If everything is OK, the output is something similar to the following:

```
BUILD SUCCESSFUL in 3s
1 actionable task: 1 executed
```

1. From a command-line terminal, move to the `Confluent` directory and start it as follows:

   ```
   $ ./bin/confluent start
   ```

2. Run a console consumer for the `uptimes` topic, shown as follows:

   ```
   $ ./bin/kafka-console-consumer --bootstrap-server localhost:9092
   --topic uptimes
   ```

3. From our IDE, run the main method of the `PlainProducer` built in previous chapters

4. The output on the console consumer of the producer should be similar to the following:

```
{"event":"HEALTH_CHECK","factory":"Lake
Anyaport","serialNumber":"EW05-
HV36","type":"WIND","status":"STARTING","lastStartedAt":"2017-09-17
T11:05:26.094+0000","temperature":62.0,"ipAddress":"15.185.195.90"}
{"event":"HEALTH_CHECK","factory":"Candelariohaven","serialNumber":
"BO58-
SB28","type":"SOLAR","status":"STARTING","lastStartedAt":"2017-08-1
6T04:00:00.179+0000","temperature":75.0,"ipAddress":"151.157.164.16
2"}
{"event":"HEALTH_CHECK","factory":"Ramonaview","serialNumber":"DV03
-
ZT93","type":"SOLAR","status":"RUNNING","lastStartedAt":"2017-07-12
T10:16:39.091+0000","temperature":70.0,"ipAddress":"173.141.90.85"}
. . .
```

5. From our IDE, run the main method of the `SparkProcessor`
6. The output on the console consumer for the `uptimes` topic should be similar to the following:

```
EW05-HV36    33
BO58-SB28    20
DV03-ZT93    46
. . .
```

Summary

If you are someone who uses Spark for batch processing, Spark Structured Streaming is a tool you should try, as its API is similar to its batch processing counterpart.

Now, if we compare Spark to Kafka for stream processing, we must remember that Spark streaming is designed to handle throughput, not latency, and it becomes very complicated to handle streams with low latency.

The Spark Kafka connector has always been a complicated issue. For example, we have to use previous versions of both, because with each new version, there are too many changes on both sides.

In Spark, the deployment model is always much more complicated than with Kafka Streams. Although Spark, Flink, and Beam can perform tasks much more complex tasks, than Kafka Streams, the easiest to learn and implement has always been Kafka.

Other Books You May Enjoy

If you enjoyed this book, you may be interested in these other books by Packt:

Building Data Streaming Applications with Apache Kafka
Manish Kumar, Chanchal Singh

ISBN: 978-1-78728-398-5

- Learn the basics of Apache Kafka from scratch
- Use the basic building blocks of a streaming application
- Design effective streaming applications with Kafka using Spark, Storm &, and Heron
- Understand the importance of a low -latency , high- throughput, and fault-tolerant messaging system
- Make effective capacity planning while deploying your Kafka Application
- Understand and implement the best security practices

Apache Kafka 1.0 Cookbook
Raúl Estrada

ISBN: 978-1-78728-684-9

- Install and configure Apache Kafka 1.0 to get optimal performance
- Create and configure Kafka Producers and Consumers
- Operate your Kafka clusters efficiently by implementing the mirroring technique
- Work with the new Confluent platform and Kafka streams, and achieve high availability with Kafka
- Monitor Kafka using tools such as Graphite and Ganglia
- Integrate Kafka with third-party tools such as Elasticsearch, Logstash, Apache Hadoop, Apache Spark, and more

Leave a review - let other readers know what you think

Please share your thoughts on this book with others by leaving a review on the site that you bought it from. If you purchased the book from Amazon, please leave us an honest review on this book's Amazon page. This is vital so that other potential readers can see and use your unbiased opinion to make purchasing decisions, we can understand what our customers think about our products, and our authors can see your feedback on the title that they have worked with Packt to create. It will only take a few minutes of your time, but is valuable to other potential customers, our authors, and Packt. Thank you!

Index

www.ingramcontent.com/pod-product-compliance
Lightning Source LLC
LaVergne TN
LVHW081526050326
832903LV00025B/1649